Air Pollution and It's Impacts on Plant Growth

NIPA® GENX ELECTRONIC RESOURCES & SOLUTIONS P. LTD.
New Delhi-110 034

Air Pollution and It's Impacts on Plant Growth

Dr Shilpa Shyam
Dr H N Verma
Dr S K Bhargava

NIPA® GENX ELECTRONIC RESOURCES & SOLUTIONS P. LTD.
New Delhi-110 034

NIPA® GENX ELECTRONIC RESOURCES & SOLUTIONS P. LTD.

101,103, Vikas Surya Plaza, CU Block
L.S.C. Market, Pitam Pura, New Delhi-110 034
Ph : +91 11 27341616, 27341717, 27341718
E-mail: newindiapublishingagency@gmail.com
www: www.nipabooks.com

For customer assistance, please contact
Phone: + 91-11-27 34 17 17
Fax: + 91-11- 27 34 16 16
E-Mail: feedbacks@nipabooks.com

ISBN: 978-81-96089-37-5

Composed and Designed by NIPA®.

PREFACE

Modern industrial society produces a large number of gases and particulate matter in sufficient quantities to harm human, animal and plant life. The amount of pollutants released in the atmosphere by fixed or mobile anthropogenic sources, is associated with economic activity. The progressive increase in urbanization in recent years, although has positive role in developmental process, but at the same time it imposes unending challenges, such as inadequate water supply, improper sanitation, and waste disposal, traffic congestion, environmental pollution and in general unsafe social environment.

Deterioration of air quality is a major environmental problem in many urban centers in both developed and developing countries. Root cause of urban air pollution is the continuous rise in vehicle population. Other causes are exponential growth of population, rapid proliferation of industries and technological development. Encroachments on roadside, mixed vehicles on roads, lack of smooth traffic flow, congested and narrow roads resulting in traffic jams are the main causes of air pollution. Urban air pollution is characterized by high concentration of suspended particulate matter, oxides of sulphur and nitrogen resulting primarily from increased use of vehicles.

Trees are immovable and are continuously exposed to pollutants in air and soil. They play an important role in maintaining ecological balance by actively participating in the cycling of nutrients, gases and air pollutants. Some trees are sensitive to pollution and act as bio-indicators of air pollution; hence it is more effective to see the impact of pollution on vegetation especially on roadside trees. Many trees are effective for trapping and absorbing air pollutants and act as sink to several air pollutants, hence trees play an important role in the improvement of urban air quality. Pollutants adversely affect plants with respect to their metabolic activities, which depend on the concentration of pollutants in the atmosphere, type of plant species, physiological stage and the season too. Plants improve air quality and thereby seem to be the suitable marker of air pollution and can be used to draw or edit air quality maps.

Contents

Chapter - 1

INTRODUCTION - URBANIZATION

Ever since industrial revolution altered the whole gamut of economic activity at least two hundred years ago, the path of economic growth including its modified form economic development had appeared unending and limitless. The whole focus of economic activity is to produce mere goods and services for use by the mankind so that life could be more comfortable and satisfying. Increasing population and increased needs of more goods and service was also seen as a hindrance or limitation in the march towards economic growth and development.

Increasing awareness about environmental degradation in the last few decades has strengthened the case of the concept of the limits to growth (Meadows *et.al.,* 1972). Environmental degradation seemed to be proportionate to growth and development. The more we try to develop, the more we degrade our environment. Intergenerational distribution of prosperity and well-being was considered to be significant aspect rather than merely the intra-generational distribution among different groups and classes. Unqualified development seemed to ensure the prosperity and well being of the present generation at the cost of future generation (Vaishampayan, 1996).

Development or growth leading to deteriorating environment in the form of worsening air quality, degraded towns and cities and damaged landscapes would make the world an unattractive location and the cities would not be attractive for investors and would cost us a great deal in economic terms as well as in our physical health and the quality of our lives.

URBANIZATION

The great innovative cities in history tend to emerge from the edge of the civilized world. They spring from the egalitarian, self-reliant societies at the moment of transition between a conservative, tradition bound past and an open oriented future. Their strength comes both from their traditions and from their new global horizons (Hall, 1994).

Cities should merge its physical/environmental planning with its social planning. The improvement of human environment goes hand in hand with the improvement of a city's use of natural resources and with the reduction of wastes. Some cities have slow traffic on streets in commercial and shopping zones to make them safer. This practice improves local air, noise traffic disturbance problems and also improves local business. Areas containing streets ranked high for pedestrian friendliness also have lower car use (Roberts, 1993).

The economic, social and environmental goals of the city need to be consolidated. The Metabolism of the City provides a way of looking at the resource inputs and waste outputs of settlements and how different types of settlements manage resource and waste (Wolman *et. al.*, 1992). Dynamics of human settlement includes population dynamics, economics, technology, management, institutional and cultural factors. Human settlement requires resource inputs like air, water, food, energy, building material and other basic materials. Livability of settlement needs health, employment, income, education, housing, accessibility, urban quality and communication. Waste outputs of the settlement are solid waste, liquid waste, air pollutants, sewage, toxics, green house gases, heat noise *etc.*

Journey of Urbanization

People throughout urban history have shown one characteristic that has shaped the nature of our cities: they do not like to travel more than half an hour to major urban destinations (Manning *et.al.,* 1980). It is possible to see that how it has caused three types of cities as transportation technologies have evolved toward greater speed and freedom of movement.

The Walking City

The traditional walking city was developed around 10,000 years ago and still exists today in some places. These cities are characterized by high population density of 100 to 200 persons per hectare and mixed land use. Such cities are joined together by narrow streets in an organic form that it's the landscape. In walking cities all destinations can be reached on foot in a short time and thus cities are rarely more than 5km. across.

The Transit City

In the latter part of the nineteenth century, old walking cities began to collapse under the pressure of population and industry. A new form of city was developed that accommodates many more people at reduced density with the maximum accessibility of half an hour through transit technology. Cities now spread to 20 to 30km and the overall density reduced to 50 to 100 persons per hectare.

The Automobile City

Beginning before the Second World War but really accelerating after it, the automobile, supplemented by bus, progressively became the transportation technology that shaped the city. It became possible to develop in any direction and low-density housing became more feasible and as a reaction to the industrial city, town planning began separating functions by zoning. The city began to decentralize and disperse. The auto city reduced in density to 10 to 20 people per hectare. The reality of most cities today is that they contain some elements of all three city types. The auto cities in developing countries are under the pressure of population rise. Continuous rise in population within a fixed area has increased the problems of automobile city

Problems of Automobile City

Assumptions, that all parts of the cities are to be easily reached from everywhere else and the city spreads beyond 40 to 50 km, have created automobile city, where the dependence on automobile is complete (Newmann and Kenworthy, 1989). Such a city will rapidly fill with traffic at most times of the day and it will show serious environmental problems together with economic and social problems.

Civilizing through technological advances is a part of the solution to the problems of automobile city, but it has increasingly been seen as insufficient because the sheer volume of vehicles overwhelms the cities.

These problems given below in a tabular form.

Environmental	**Economic**	**Social**
Oil Vulnerability	External cost from accidents, pollution, health impacts *etc.*	Loss of street life
Photochemical smog	Congestion costs, despite endless road building	Loss of community
Air pollution	High infrastructure costs in new sprawling suburbs	Loss of public safety
High green house gas contribution	Loss of productive rural land	Isolation in remote suburbs
Greater storm water problems from extra surface	Loss of urban land to bitumen	Access problems for the careless and those with disabilities
Traffic problems-noise, severance	Loss of time due to sprawl increasing distances	

Cities In Developing Countries

In most of the developing countries, while the fast growing urbanization has positive role in the progress of economic development but also it poses unending challenges such as shortage of housing, inadequate and failure of water supply, improper sanitation, and waste disposal facilities, congestion, traffic problems, environmental pollution and in general an unsafe social environment. Turning of old age agriculture base society into industry business and job-oriented society, leading to migration of more and more villagers to the city, as a result cities are over crowded loosing its good health and carrying capacity. Exponential growth of population, rapid proliferation of industrialization and technological development, moreover there have been haphazard and awkward planning or semi planned growth of cities and encroachments on roadsides, concentration of offices cum commercial places, mixed vehicles plying on the roads, lack of smooth traffic flow and congested and narrow roads are root causes of traffic jams resulting in more air pollution.

Traditional	Transit City pre modern walking city	Automobile	Future city	Urbanism
Economy and technology	Small household industries *i.e.* local and small regional economy	Latter industries concentrated in parts of cities *i.e.* national and regional economy	Large scale industries scattered *i.e.* national and regional economy	Information and services oriented *i.e.* global economy, industries move to rural areas and towns
Social organization	Person to person community	Bigger cities began losing person to person contact but still community oriented in rail based suburbs	Individualistic and isolated	Local and Global communication
Transportation	Walking and cycling latter	Trams and trains with walking	Automobiles	Walking and cycling (local, city), vehicles (supplementary), air (global)
Urban form	Small, dense mixed and organic	Medium density dense mixed centers and corridors with green wedges	High rise buildings, low suburbs and high density centers	Local urban villages (high across city) by transit medium and low density areas around villages, no more sprawl
Environment	Low resources, low wastes, close to rural areas	Medium resources and wastes	High resources and wastes, little nature oriented	Low resources low wastes, close to nature in local spaces, regional and global orientation

Newmann and Kenworthy, 1989, gave land use pattern, transportation and environmental conditions of different types of cities.

Cutting of trees for housing and fuel is increasing. As a result urban environmental pollution is rising manifolds, which ultimately leads to the overall degradation of environment (air, water, soil and vegetation).

Chapter - 2

AIR POLLUTION

AIR

Air is a non-homogenous mixture of gases, solid particlcs, and liquids; It consists of gases and particles with small settling velocity that exhibit stability in a gravitational field. Atmosphere has four major layers Troposphere from 0-17km., Stratosphere from 17-50km., Mesosphere from 50-90 km. and Ionosphere is from 90-100 km. (95% of the air by weight is contained in the Troposphere *i.e.* from 0-17 km).

Major components of air are (99%) in

Component	**% by wt. or by fraction of total molecules***
N_2	75.51
O_2	23.14
Ar	1.28
Water vapor	

* Percentage by weight and *minor components* are

Component	**ppm****
CO_2	325.0
Ne	18.0
He	5.0
CH_4	2.0
Kr	1.0
H_2	0.5
N_2O	0.5
Xe	0.1

** parts per million, 325 ppm means that 325 of each 1 million particle is CO_2

Air Pollution

Clean air is a basic requirement for human. Vehicles pouring forth black fumes are common phenomena in cities of developing countries. One may fall victim to much less visible health hazards from the air we breathe even in urban areas where air pollution control measures were first put in place nearly thirty years ago.

Air pollution was first perceived as a local problem in urban industrialized area hence, it was thought that diluting whatever was put into the atmosphere would sufficiently ameliorate potential problems. Thus increased height of smoke stacks to allow wind to disperse pollutants over wide areas was a ready solution. However taller stacks merely transported the problem elsewhere and soon regional problem such as acid rain was recognized. More recently global problems such as climate change and stratospheric ozone depletion were identified and publicized. Hence laws and regulations to correct and control local, regional and global environmental problems. Air pollution measures require cooperation from city ordinances to state laws and finally requiring full participation in international agreements because, air is a resource that respects no man made boundaries. The approach thinking globally about air quality can be emphasized by the following inter links between air pollutants and their host medium the atmosphere:

» *Air pollutants* are inseparably associated with changes in earth's atmosphere

» The various air pollution problems occurring at a local scale involve many of the pollutants that also cause problems globally and the problems are interconnected

» Secondary pollutant formation of various species occurs under short and long-range transport processes.

Clean air can no longer be taken for granted. Today, the air in most large Indian cities is severely polluted and this pollution has tremendous impact on health of humans as well as of plants. Population, industrialization and growth in number of vehicles in urban areas has lead to a rapid deterioration of outdoor air quality. Out of 23 mega and metro cities in our country, Delhi is the most polluted followed by Mumbai, Kolkata, Banglore, Chennai, Kanpur, Ahmedabad, and Nagpur (CPCB, 1999).

Causes Of Air Pollution

Under the process of industrialization, events like expansion of cities, increase in traffic, rapid economic development and higher levels of energy consumption cause air pollution. In developing countries, the growth of cities is non-planned/structured/zoned thus increasing the air pollution problems. Combustion of fossil fuel by stationary (power plants and other industries) and mobile sources (motor vehicles) leads to air pollution in urban centers (Yunus *et.al.,* 1996) and root cause of air pollution in Indian cities is vehicular emissions (fuel cell bus development in India, 2000).

Major causes of air pollution are:

Population

The *world's population* increased dramatically after 1950 and over population has been a problem ever since. According to the international Programs Center, U.S. Bureau of Census, the total population of the World as on May 15, 2005 was 6,441,601,382 at 11:24 GMT. Most high concentration of any type of pollution takes place in big cities with millions of peoples. Los Angeles, Washington and New York cities have population density of 20, 13 and 20 persons /hectare. European cities (Stockholm and Amsterdam) have population density of 51 persons /hectare, Australian cities (Melbourne, Perth and Sydney) have 16, 11, 18 persons /hectare respectively) and Asian cities (Tokyo) have population density of 105 persons/hectare (Newman and Kenworthy1989).

The number of cities with more than 7 million inhabitants increased from 13 to 35 between 1950 and 1980 (UN, 1989) and majority of these cities were in developing countries. Land area of US was 35% higher than that of India whereas population of India was 3.6 times of the total population of US. Hence the developing countries are under the pressure of population. According to WHO/UNEP report (1992), if current trend continues Mexico, Sao Paulo, and Tokyo will be three largest cities in the world having over 50 million inhabitants each by the year 2000.

A little more than 4 years ago, in 1999, global population surpassed 6 billion. At midyear 2002, it stood at 6.2 billion and just over two people were being added each second. As rapid as this may seem, the pace at which global population was growing had already peaked more than a decade earlier. In absolute terms, approximately 74 million people were added to the world's population in 2002 compared to a high of 87 million in 1989-90. Similarly, the annual average growth rate was approximately 1.2 percent in 2002, down from a high of 2.2 percent in 1963-64. It is expected that this slowdown in population growth will continue into the foreseeable future. The slowdown in the growth of the world's population can be traced primarily to decline in fertility. In 2002, the world's women, on average, were giving birth to 2.6 children over their lifetime. This was less than one-half of a child more than the level needed to assure the replacement of the population. Although fertility rates in some parts of the world are expected to remain above replacement level for quite some time (e.g., in Sub- Saharan Africa) (Global population profile, U.S. Department of Commerce Economics and Statistics Administration U.S. CENSUS BUREAU Issued March 2004).

According to the *CIA- The world fact book* 2004 population and areas of countries is given below:

Country	Abbr.	Status	Admn.	Capital	Area km^2	Population
Afghanistan	AFG	Indep		Käbol	647,500	28,513,700
Albania	ALB	Indep		Tiranë	28,748	3,544,800
Algeria	DZA	Indep		El Djazaïr	2,381,740	32,129,300
American Samoa	ASM	DepT	USA	Pago Pago	199	57,900
Andorra	AND	Indep		Andorra la Vella	468	69,900
Angola	AGO	Indep		Luanda	1,246,700	10,978,600
Anguilla	AIA	DepT	GBR	The Valley	102	13,000
Antigua and Barbuda	ATG	Indep		Saint John's	443	68,300
Argentina	ARG	Indep		Buenos Aires	2,766,890	39,144,800
Armenia	ARM	Indep		Jerevan	29,800	2,991,400
Aruba	ABW	DepT	NLD	Oranjestad	193	71,200
Australia	AUS	Indep		Canberra	7,686,850	19,913,100
Austria	AUT	Indep		Wien	83,858	8,174,800
Azerbaijan	AZE	Indep		Baký‌	86,600	7,868,400
Bahamas	BHS	Indep		Nassau	13,940	299,700
Bahrain	BHR	Indep		Al-Manämah	665	677,900

Country	Abbr.	Status	Admn.	Capital	Area km²	Population
Bangladesh	BGD	Indep		Dhaka	144,000	41,340,500
Barbados	BRB	Indep		Bridgetown	431	278,300
Belarus	BLR	Indep		Minsk	207,600	10,310,500
Belgium	BEL	Indep		Bruxelles	30,510	10,348,300
Belize	BLZ	Indep		Belmopan	22,966	272,900
Benin	BEN	Indep		Porto-Novo / Cotonou	112,620	7,250,000
Bermuda	BMU	DepT	GBR	Hamilton	53	64,900
Bhutan	BTN	Indep		Thimphu	47,000	2,185,600
Bolivia	BOL	Indep		Sucre/La Paz	1,098,580	8,724,200
Bosnia and Herzegovina	BIH	Indep		Sarajevo	51,129	4,007,600
Botswana	BWA	Indep		Gaborone	600,370	1,562,000
Brazil	BRA	Indep		Brasília	8,511,965	84,101,100
British Virgin Islands	VGB	DepT	GBR	Road Town	153	22,200
Brunei	BRN	Indep		Bandar Seri Begawan	5,770	365,300
Bulgaria	BGR	Indep		Sofija	110,910	7,518,000
Burkina Faso	BFA	Indep		Ouagadougou	274,200	13,574,800
Burundi	BDI	Indep		Bujumbura	27,830	6,231,200
Cambodia	KHM	Indep		Phnum Pénh	181,040	13,363,400
Cameroon	CMR	Indep		Yaoundé	475,440	16,063,700
Canada	CAN	Indep		Ottawa	9,976,140	32,507,900
Cape Verde	CPV	Indep		Praia	4,033	415,300
Cayman Islands	CYM	DepT	GBR	George Town	262	43,100
Central African Republic	CAF	Indep		Bangui	622,984	3,742,500
Chad	TCD	Indep		Ndjamena	1,284,000	9,538,500
Chile	CHL	Indep		Santiago / Valparaíso	756,950	15,824,000
China	CHN	Indep		Beijing	9,596,960	1,298,847,600
Colombia	COL	Indep		Bogotá	1,138,910	42,310,800
Comoros	COM	Indep		Moroni	2,170	651,900
Congo (Dem. Rep.)	COD	Indep		Kinshasa	2,345,410	58,317,900
Congo (Rep.)	COG	Indep		Brazzaville	342,000	2,998,000

Country	Abbr.	Status	Admn.	Capital	Area km²	Population
Cook Islands	COK	DepT	NZL	Avarua	240	21,200
Costa Rica	CRI	Indep		San José	51,100	3,956,500
Cote d'Ivoire	CIV	Indep		Yamoussoukro / Abidjan	322,460	17,327,700
Croatia	HRV	Indep		Zagreb	56,542	4,496,900
Cuba	CUB	Indep		La Habana	110,860	11,308,800
Cyprus	CYP	Indep		Levkosía	9,250	775,900
Czech Republic	CZE	Indep		Praha	78,866	10,246,200
Denmark	DNK	Indep		København	43,094	5,413,400
Djibouti	DJI	Indep		Djibouti	23,000	466,900
Dominica	DMA	Indep		Roseau	754	69,300
Dominican Republic	DOM	Indep		Santo Domingo	48,730	8,833,600
East Timor	TMP	Indep		Dili	15,007	1,019,300
Ecuador	ECU	Indep		Quito	283,560	13,212,700
Egypt	EGY	Indep		Al-Qähirah	1,001,450	76,117,400
El Salvador	SLV	Indep		San Salvador	21,040	6,587,500
Equatorial Guinea	GNQ	Indep		Malabo	28,051	523,100
Eritrea	ERI	Indep		Asmera	121,320	4,447,300
Estonia	EST	Indep		Tallinn	45,226	1,341,700
Ethiopia	ETH	Indep		Adis Abeba	1,127,127	67,851,300
Falkland Islands	FLK	DepT	GBR	Stanley	12,173	2,970
Faroe Islands	FRO	DepT	DNK	Tórshavn	1,399	46,700
Fiji	FJI	Indep		Suva	18,270	880,900
Finland	FIN	Indep		Helsinki	337,030	5,214,500
France	FRA	Indep		Paris	547,030	60,424,200
French Guiana	GUF	DepT	FRA	Cayenne	91,000	191,300
French Polynesia	PYF	DepT	FRA	Papeete	4,167	266,300
Gabon	GAB	Indep		Libreville	267,667	1,355,200
Gambia	GMB	Indep		Banjul	11,300	1,546,800
Georgia	GEO	Indep		Tbilisi	69,700	4,693,900
Germany	DEU	Indep		Berlin	357,021	82,424,600
Ghana	GHA	Indep		Accra	239,460	20,757,000

Country	Abbr.	Status	Admn.	Capital	Area km²	Population
Gibraltar	GIB	DepT	GBR	Gibraltar	6	27,800
Great Britain (and Northern Ireland)	GBR	Indep		London	244,820	60,270,700
Greece	GRC	Indep		Athínai	131,940	10,647,500
Greenland	GRL	DepT	DNK	Nuuk	2,166,086	56,400
Grenada	GRD	Indep		Saint George's	344	89,400
Guadeloupe	GLP	DepT	FRA	Basse-Terre	1,780	444,500
Guam	GUM	DepT	USA	Agana	549	166,100
Guatemala	GTM	Indep		Ciudad de Guatemala	108,890	14,280,600
Guernsey	GUE	DepT	GBR	St. Peter Port	78	65,000
Guinea	GIN	Indep		Conakry	245,857	9,246,500
Guinea-Bissau	GNB	Indep		Bissau	36,120	1,388,400
Guyana	GUY	Indep		Georgetown	214,970	705,800
Haiti	HTI	Indep		Port-au-Prince	27,750	7,656,200
Honduras	HND	Indep		Tegucigalpa	112,090	6,823,600
Hong Kong	HKG	DepT	CHN	Victoria	1,092	6,855,100
Hungary	HUN	Indep		Budapest	93,030	10,032,400
Iceland	ISL	Indep		Reykjavík	103,000	294,000
India	IND	Indep		New Delhi	3,287,590	1,065,070,600
Indonesia	IDN	Indep		Jakarta	1,919,440	238,453,000
Iran	IRN	Indep		Tehrân	1,648,000	69,018,900
Iraq	IRQ	Indep		Baghdâd	437,072	25,374,700
Ireland	IRL	Indep		Dublin	70,280	3,969,600
Israel	ISR	Indep		Yerushalayim	20,770	6,199,000
Italy	ITA	Indep		Roma	301,230	58,057,500
Jamaica	JAM	Indep		Kingston	10,991	2,713,100
Japan	JPN	Indep		Tôkyô	377,835	127,333,000
Jersey	JER	DepT	GBR	St. Helier	116	90,500
Jordan	JOR	Indep		Ammân	92,300	5,611,200
Kazakhstan	KAZ	Indep		Astana	2,717,300	15,143,700
Kenya	KEN	Indep		Nairobi	582,650	32,021,900
Kiribati	KIR	Indep		Tarawa	811	100,800
Korea (North)	PRK	Indep		P'yongyang	120,540	22,697,600
Korea (South)	KOR	Indep		Soul	98,480	48,598,200

Country	Abbr.	Status	Admn.	Capital	Area km²	Population
Kuwait	KWT	Indep		Al-Kuwayt	17,820	2,257,500
Kyrgyzstan	KGZ	Indep		Biškek	198,500	5,081,400
Laos	LAO	Indep		Viangchan	236,800	6,068,100
Latvia	LVA	Indep		Rîga	64,589	2,306,300
Lebanon	LBN	Indep		Bayrût	10,400	3,777,200
Lesotho	LSO	Indep		Maseru	30,355	1,865,000
Liberia	LBR	Indep		Monrovia	111,370	3,390,600
Libya	LBY	Indep		Tarâbulus	1,759,540	5,631,600.
Liechtenstein	LIE	Indep		Vaduz	160	33,400
Lithuania	LTU	Indep		Vilnius	65,200	3,607,900
Luxembourg	LUX	Indep		Luxembourg	2,586	462,700
Macau	MAC	DepT	CHN	Macau	25	445,300
Macedonia	MKD	Indep		Skopje	25,333	2,071,200
Madagascar	MDG	Indep		Antananarivo	587,040	17,501,900
Malawi	MWI	Indep		Lilongwe	118,480	11,906,900
Malaysia	MYS	Indep		Kuala Lumpur	329,750	23,522,500
Maldives	MDV	Indep		Male	300	339,300
Mali	MLI	Indep		Bamako	1,240,000	11,956,800
Malta	MLT	Indep		Valletta	316	396,900
Man, Isle of	MAN	DepT	GBR	Douglas	572	74,700
Marshall Islands	MHL	Indep		Delap-Uliga-Darrit	181	57,700
Martinique	MTQ	DepT	FRA	Fort-de-France	1,100	429,500
Mauritania	MRT	Indep		Nouakchott	1,030,700	2,998,600
Mauritius	MUS	Indep		Port Louis	2,040	1,220,500
Mayotte	MYT	DepT	FRA	Dzaoudzi	374	186,000
Mexico	MEX	Indep		Ciudad de México	1,972,550	104,959,600
Micronesia, Federated States of	FSM	Indep		Palikir	702	108,200
Moldova	MDA	Indep		Chi°inău	33,843	4,446,500
Monaco	MCO	Indep		Monaco	2	32,300
Mongolia	MNG	Indep		Ulaanbaatar	1,565,000	2,751,300
Montserrat	MSR	DepT	GBR	Plymouth	102	9,250
Morocco	MAR	Indep		Rabat	446,550	32,209,100
Mozambique	MOZ	Indep		Maputo	801,590	18,811,700

Country	Abbr.	Status	Admn.	Capital	Area km²	Population
Myanmar	MMR	Indep		Yangon	678,500	42,720,200
Namibia	NAM	Indep		Windhoek	825,418	1,954,000
Nauru	NRU	Indep		Yaren	21	12,800
Nepal	NPL	Indep		Kâthmândau	140,800	27,070,700
Netherlands	NLD	Indep		Amsterdam / 's-Gravenhage	41,526	16,318,200
Netherlands Antilles	ANT	DepT	NLD	Willemstad	960	218,100
New Caledonia	NCL	DepT	FRA	Nouméa	19,060	213,700
New Zealand	NZL	Indep		Wellington	268,680	3,993,800
Nicaragua	NIC	Indep		Managua	129,494	5,359,800
Niger	NER	Indep		Niamey	1,267,000	11,360,500
Nigeria	NGA	Indep		Abuja	923,768	137,253,100
Niue	NIU	DepT	NZL	Alofi	260	2,160
Norfolk Island	NFK	DepT	AUS	Kingston	35	1,840
Northern Mariana Islands	MNP	DepT	USA	Garapan	477	78,300
Norway	NOR	Indep		Oslo	324,220	4,574,600
Oman	OMN	Indep		Masqat	212,460	2,903,200
Pakistan	PAK	Indep		Islâmâbâd	803,940	159,196,300
Palau	PLW	Indep		Koror	458	20,000
Palestinian Territories	PLT				6,220	3,636,200
Panama	PAN	Indep		Ciudad de Panamá	78,200	3,000,500
Papua New Guinea	PNG	Indep		Port Moresby	462,840	5,420,300
Paraguay	PRY	Indep		Asunción	406,750	6,191,400
Peru	PER	Indep		Lima	1,285,220	27,544,300
Philippines	PHL	Indep		Manila	300,000	86,241,700
Pitcairn	PCN	DepT	GBR	Adamstown	47	46
Poland	POL	Indep		Warszawa	312,685	38,626,300
Portugal	PRT	Indep		Lisboa	92,391	10,524,100
Puerto Rico	PRI	DepT	USA	San Juan	9,104	3,898,000
Qatar	QAT	Indep		Ad-Dawhah	11,437	840,300
Reunion	REU	DepT	FRA	Saint-Denis	2,517	766,200
Romania	ROM	Indep		Bucure°ti	237,500	22,355,600
Russia	RUS	Indep		Moskva	17,075,200	143,782,300
Rwanda	RWA	Indep		Kigali	26,338	7,954,000

Country	Abbr.	Status	Admn.	Capital	Area km²	Population
Saint Helena	SHN	DepT	GBR	Jamestown	410	7,420
Saint Kitts and Nevis	KNA	Indep		Basseterre	261	38,800
Saint Lucia	LCA	Indep		Castries	616	164,200
Saint Pierre and Miquelon	SPM	DepT	FRA	Saint-Pierre	242	7,000
Saint Vincent and the Grenadines	VCT	Indep		Kingstown	389	117,200
Samoa	WSM	Indep		Apia	2,944	177,700
San Marino	SMR	Indep		San Marino	61	28,500
Sao Tome and Principe	STP	Indep		São Tomé	1,001	181,600
Saudi Arabia	SAU	Indep		Ar-Riyâd	1,960,582	25,795,900
Senegal	SEN	Indep		Dakar	196,190	10,852,100
Serbia and Montenegro	YUG	Indep		Beograd	102,350	10,825,900
Seychelles	SYC	Indep		Victoria	455	80,800
Sierra Leone	SLE	Indep		Freetown	71,740	5,883,900
Singapore	SGP	Indep		Singapore	693	4,353,900
Slovakia	SVK	Indep		Bratislava	48,845	5,423,600
Slovenia	SVN	Indep		Ljubljana	20,273	2,011,500
Solomon Islands	SLB	Indep		Honiara	28,450	523,600
Somalia	SOM	Indep		Muqdisho	637,657	8,304,600
South Africa	ZAF	Indep		Pretoria / Cape Town	1,219,912	42,718,500
Spain	ESP	Indep		Madrid	504,782	40,280,800
Sri Lanka	LKA	Indep		Colombo	65,610	19,905,200
Sudan	SDN	Indep		Al-Khartûm	2,505,810	39,148,200
Suriname	SUR	Indep		Paramaribo	163,270	436,900
Svalbard	SJM	DepT	NOR	Longyearbyen	62,049	2,760
Swaziland	SWZ	Indep		Mbabane	17,363	1,169,200
Sweden	SWE	Indep		Stockholm	449,964	8,986,400
Switzerland	CHE	Indep		Bern	41,290	7,450,900
Syria	SYR	Indep		Dimashq	185,180	18,016,900
Taiwan (Republic of China)	TWN	Indep		Taipei	35,980	22,749,800

Country	Abbr.	Status	Admn.	Capital	Area km²	Population
Tajikistan	TJK	Indep		Dushanbe	143,100	7,011,600
Tanzania	TZA	Indep		Dodoma / Dar es Salaam	945,087	36,588,200
Thailand	THA	Indep		Krung Thep	514,000	64,865,500
Togo	TGO	Indep		Lomé	56,785	5,556,800
Tokelau	TKL	DepT	NZL		10	1,410
Tonga	TON	Indep		Nuku'alofa	748	110,200
Trinidad and Tobago	TTO	Indep		Port-of-Spain	5,128	1,096,600
Tunisia	TUN	Indep		Tûnis	163,610	9,974,700
Turkey	TUR	Indep		Ankara	780,580	68,893,900
Turkmenistan	TKM	Indep		Ashgabad	488,100	4,863,200
Turks and Caicos Islands	TCA	DepT	GBR	Cockburn Town	430	20,000
Tuvalu	TUV	Indep		Vaiaku	26	11,500
Uganda	UGA	Indep		Kampala	236,040	26,404,500
Ukraine	UKR	Indep		Kyyiv	603,700	47,732,100
United Arab Emirates	ARE	Indep		Abû Zaby	82,880	2,523,900
United States of America	USA	Indep		Washington	9,629,091	293,027,600
Uruguay	URY	Indep		Montevideo	176,220	3,399,200
Uzbekistan	UZB	Indep		Toshkent	447,400	26,410,400
Vanuatu	VUT	Indep		Port Vila	12,200	202,600
Vatican City	VAT	Indep			1	920
Venezuela	VEN	Indep		Caracas	912,050	25,017,400
Vietnam	VNM	Indep		Ha Noi	329,560	82,689,500
Virgin Islands of the US	VIR	DepT	USA	Charlotte Amalie	352	108,800
Wallis and Futuna	WLF	DepT	FRA	Mata-Utu	274	15,900
Western Sahara	ESH			Al-'Ayun	266,000	267,400
Yemen	YEM	Indep		San'â'	527,970	20,024,900
Zambia	ZWB	Indep		Lusaka	752,614	10,462,400
Zimbabwe	ZWE	Indep		Harare	390,580	12,671,900
World					...	**6,379,157,000**

Source: http://www.citypopulation.de

The total *population of India* as at 0:00 hours on 1st March 2001 stood at 1,027,015,247 persons (Census of India 2001) with a decadal growth of (+21.34%). With this India became only the second country in the world after China to cross the one billion mark. The population of the country rose by 21.34% during 1991-2001. Total urban population in India is approximately 30% of total population (fuel cell bus development in India, 2000). The population growth has been sudden and spectacular in India. In 1901 Delhi had a population of barely four Lakhs, and 26.58 Lakhs in 1961. But by 1991 the population of Delhi had soared to 94.20 Lakhs, forming 1.11% of the country's population (MCD, Delhi, 1999). According to the Census of India 2001, state wise population of India is given below:

India/States Union Territories*	Population			Population Variation 1991-2001	Sex ratio (females per thousand males)
	Persons	Males	Females		
INDIA 1,2	1,027,015,247	531,277,078	495,738,169	21.34	933
Andaman & Nicobar Is.*	356,265	192,985	163,280	26.94	846
Andhra Pradesh	75,727,541	38,286,811	37,440,730	13.86	978
Arunachal Pradesh	1,091,117	573,951	517,166	26.21	901
Assam	26,638,407	13,787,799	12,850,608	18.85	932
Bihar	82,878,796	43,153,964	39,724,832	28.43	921
Chandigargh*	900,914	508,224	392,690	40.33	773
Chhatisgarh	20,795,956	10,452,426	10,343,530	18.06	990
Dadra & Nagar Haveli*	220,451	121,731	98,720	59.2	811
Daman & Diu*	158,059	92,478	65,581	55.59	709
Delhi*	13,782,976	7,570,890	6,212,086	46.31	821
Goa	1,343,998	685,617	658,381	14.89	960
Gujarat 5	50,596,992	26,344,053	24,252,939	22.48	921
Haryana	21,082,989	11,327,658	9,755,331	28.06	861
Himachal Pradesh 4	6,077,248	3,085,256	2,991,992	17.53	970
Jammu & Kashmir 2,3	10,069,917	5,300,574	4,769,343	29.04	900
Jharkhand	26,909,428	13,861,277	13,048,151	23.19	941
Karnataka	52,733,958	26,856,343	25,877,615	17.25	964

India/State/ Union Territories*	Population			Population Variation 1991-2001	Sex ratio (females per thousand males)
	Persons	Males	Females		
Kerala	31,838,619	15,468,664	16,369,955	9.42	1,058
Lakshadweep*	60,595	31,118	29,477	17.19	947
Madhya Pradesh	60,385,118	31,456,873	28,928,245	24.34	920
Maharashtra	96,752,247	50,334,270	46,417,977	22.57	922
Manipur	2,388,634	1,207,338	1,181,296	30.02	978
Meghalaya	2,306,069	1,167,840	1,138,229	29.94	975
Mizoram	891,058	459,783	431,275	29.18	938
Nagaland	1,988,636	1,041,686	946,950	64.41	909
Orissa	36,706,920	18,612,340	18,094,580	15.94	972
Pondicherry*	973,829	486,705	487,124	20.56	1,001
Punjab	24,289,296	12,963,362	11,325,934	19.76	874
Rajasthan	56,473,122	29,381,657	27,091,465	28.33	922
Sikkim	540,493	288,217	252,276	32.98	875
Tamil Nadu	62,110,839	31,268,654	30,842,185	11.19	986
Tripura	3,191,168	1,636,138	1,555,030	15.74	950
Uttar Pradesh	166,052,859	87,466,301	78,586,558	25.8	898
Uttaranchal	8,479,562	4,316,401	4,163,161	19.2	964
West Bengal	80,221,171	41,487,694	38,733,477	17.84	934

Notes:

1. The population of India includes the estimated population of entire Kachchh district, Morvi, Maliya-Miyana and Wankaner talukas of Rajkot district, Jodiya taluka of Jamanagar district of Gujarat State and entire Kinnaur district of Himachal Pradesh where population enumeration of Census of India 2001 could not be conducted due to natural calamity.
2. For working out density of India, the entire area and population of those portions of Jammu and Kashmir which are under illegal occupation of Pakistan and China have not been taken into account.
3. Figures shown against Population in the age-group 0-6 and Literates do not include the figures of entire Kachchh district, Morvi, Maliya-Miyana and Wankaner talukas of Rajkot district, Jodiya taluka of Jamanagar district and entire Kinnaur district of Himachal Pradesh where population enumeration of Census of India 2001 could not be conducted due to natural calamity.
4. Figures shown against Himachal Pradesh have been arrived at after including the estimated figures of entire Kinnaur district of Himachal Pradesh where the

population enumeration of Census of India 2001 could not be conducted due to natural calamity.

5. Figures shown against Gujarat have been arrived at after including the estimated figures of entire Kachchh district, Morvi, Maliya-Miyana and Wankaner talukas of Rajkot district, Jodiya taluka of Jamnagar district of Gujarat State where the population enumeration of Census of India 2001 could not be conducted due to natural calamity.

(*Source:* Provisional Population Totals : India . Census of India 2001, Paper 1 of 2001)

Transportation

Cities are mostly shaped by decision and priorities concerning transportation. The energy component of transportation is a critical part of the metabolism of the city (Newman and Kenworthy, 1989). US cities use on average 58.5 GJ of fuel per capita for transportation energy compared to 34.7 GJ per capita in Australian cities, 13.3 GJ in European cities, and 5.5 GJ in Asian cities. This is an enormous variation, much greater than could be explained by obvious economic factors. In most developing countries the transportation sector accounts for one third of total commercial energy consumption and more than one half of total oil consumption. China and India are exceptions, in these countries transportation accounts for less than 10 percent of commercial energy consumption, though this share is growing rapidly because of economic development and increase in vehicle population. (Faiz 1996).

Vehicle Population

Use of motor vehicles is growing fast in developing countries. Since 1979, the greatest increase has been in Asia. Motor vehicles contributions are less important in cities with lower levels of motorization and in cities located where strong control measures have been established, *e.g.,* in temperate regions. In many developing countries vehicle fleet tend to be older and poorly maintained, thus increasing the significance of motor vehicle as a pollutant source. The number of vehicles is expected to grow faster in the developing countries than in the developed countries. The share of motor vehicles in the pollution load is thus set to increase in developing countries and in the absence of the introduction of stringent control measures for traffic related pollutants, air quality will deteriorate further in developing countries (Yunus and Iqbal, 1996).

Automobiles have now become an integral necessity for the management of modern life but their private and commercial uses have come to dominate the whole world. In 1930, there were about 37 million globally registered vehicles, which has grown to 675 million in 1990 and rose upto 754 million in 2000. It is expected that by 2020 the vehicle population will cross 1,116,000 (Motor Vehicle Manufacturers, 1991). In 2000 total vehicle population of developed and developing countries was 538 million and 216 million (Pemberton, Max., 2000). Vehicle population includes passenger and commercial vehicles. More developed countries includes Australia, Austria, Belgium, Canada, Finland, France, Germany, Greece, Iceland, Irish Republic, Italy, Japan, Luxembourg, Mexico, Netherlands, New Zeland, Norway, Portugal, Spain, Sweden, UK and USA. Developing countries include all others.

Motorcar has become an indispensable part of modern life. There is an increasing concern about its environmental impact, particularly the negative effects of automotive exhaust emissions on air quality and human health. It has long been acknowledged that pollutants emitted from gasoline-driven vehicles contribute to a decline in air quality. Carbon monoxide, nitrogen oxide and hydrocarbons are all known to be toxic to humans or damaging to the environment.

The proportion of pollutants in the air, which are directly attributable to vehicle emissions, has risen significantly in line with the growth in car ownership and the increase in the number of kilometers driven per car. Total annual car sale of world in 1997-98 was 41.57 million. US has the highest car sale (8.90 million) followed by Japan (8.49 million), Germany (4.68 million) and France (3.35million), whereas the total annual car sale of India in 1997-98 was 0.41 million (AID Newsletter / Facts & Figures, 1997-98).

Numerous studies have been undertaken in Europe to determine the impact of car emissions on human health and the environment. The results are alarming. In Europe, the 1999, WHO report on Health costs due to road traffic-related air pollution revealed that car-related pollution kills more people than car accidents in the three European countries where the study took place (Austria, France and Switzerland). The main findings of this report were:

» Long-term exposure to air pollution from cars in adults over 30 years of age caused an extra 21,000 premature deaths per year from respiratory or heart disease. This is more than the total annual deaths from road traffic accidents in the countries studied (9,900).

» Each year, air pollution from cars causes 300,000 extra cases of bronchitis in children, plus 15,000 hospital admissions for heart disease, 395,000 asthma attacks in adults and 162,000 attacks in children.

Vehicle population of India

Vehicle population in India has also grown very rapidly in the last decade. Two and three wheelers have multiplied by a factor of five while the passenger cars and diesel vehicles have increased by more than two and half times during this period. Vehicle population in India was estimated to be 27.8 million in 1994-95 and is currently (2000-2001) estimated at 53.0 million. During the next 5 years the vehicle population may grow to more than 75 million. Two & three wheelers account for more than two-third of total vehicle population. According to the GITE Regional Workshop on Inspection and Maintenance Policy in Asia Entitled Inspection & Maintenance for In-use Vehicles in India, 10-12 December 2001, Bangkok, Thailand. The growth of vehicle population in India is given below:

Vehicle Population (in Millions)

Year	2-W	3-W	Cars	MUVs	CVs	Total
1986	6.13	0.39	1.28	0.39	1.06	9.25
1991	14.05	0.76	2.28	0.73	1.74	19.56
1998	28.74	1.40	3.94	0.86	2.31	37.25
2000	38.77	2.51	4.72	1.10	2.39	49.49

MUVs- Multi-utility vehicles
CVs- Commercial vehicles

The increase in the number of vehicles on the roads of India has resulted in the increased traffic congestions in the mega cities and consequently a substantial increase in pollution levels. Pollution from vehicles increased from 57% in 1975 to 74% in 1995 as per Citizens

Fifth Report, 2000. Six metropolitan cities, Bangalore, Calcutta, Chennai, Delhi, Hyderabad, and Mumbai account for almost one third of the total vehicle population in India. Total vehicle population of India in 1998 was 4.09 crore. It includes two wheelers (2.83 crore), car, jeep and taxis (50 lakhs), Buses (5.3 lakhs) goods Vehicles (25 lakhs) and others (44 lakhs) (Ministry of Surface Transport 1999).

Total registered population of *two stroke two wheelers, two stroke three wheelers, four stroke two wheelers and four stroke three wheelers* in India (2000) was 320.14, 16.78, 42.84 and 3.94 lakhs respectively. Total population of two wheelers and three wheelers in 2000 was 36.268 and 20.72 lakhs respectively (N.V. Iyer, Bajaj Auto Ltd., Pune 2000). Data indicates that the number of vehicles operated by two stroke engine is higher than vehicles operated by four stroke engines. Cars in Mumbai, Delhi, Banglore and Chennai as Percentage of Total Vehicle on Indian city roads were 32, 24, 14 and 17% respectively (Motor Transport Statistics of India, 2000).

Delhi recorded the largest vehicle population in 1995 among the major cities in India. Two stroke two wheelers in 1996-2000 showed a 7.74% rise and four stroke two wheelers showed a 3.3 times increase in comparison to 1991-1995. Gasoline fueled passenger cars and Multi utility vehicles showed a rise of 64.86% and 86.63% respectively in 1996-2000. Diesel fueled passenger cars showed a 15 times rise in Delhi during 1996-2000 (CPCB, 2000). The population of different kinds of vehicles in Delhi during the period of 1993-97 and during the period 1996-2000 for Mumbai is given below:

Number of registered motor vehicle in Delhi

Type of vehicle	1993	1994	1995	1996	1997
Cars & Jeep	510242	557543	617585	685850	705923
Motorcycle & Scooters (two wheelers)	1467182	1580817	1707528	1844471	1876053
Auto Rickshaw (three wheelers)	71568	74408	77884	80208	80210
Taxis	11679	12225	13384	14593	15015
Buses	23943	25553	27473	29183	29572
Goods vehicle *etc.*	114294	122444	131877	139300	140922
Total	**2198908**	**2372990**	**2575731**	**2793605**	**2847695**

Motor Vehicle Population in Mumbai

Vehicle Category	1996	1997	1998	1999	2000
Motor cycle	1,29,372	1,41,535	1,53,435	1,66,649	1,83,972
Scooters	1,46,811	—	—	1,82,289	1,91,541
Mopeds	26,330	—	—	30,503	31,793
Total 2 Wheelers	3,02,513	3,28,940	3,54,799	3,79,441	4,07,306
Three Wheelers / Auto rickshaws	59,222	72,007	83,705	91,622	97,565
Cars / Jeeps / Station Wagons	2,52,698	2,79,613	2,98,905	3,10,943	3,29,546
Taxis	44,842	48,646	51,959	55,472	58,696
Buses	12,277	12,809	13,138	13,555	14,096
Good Vehicles	44,684	48,206	50,072	52,184	53,980
Tractors	1,120	1,180	1,225	1,270	1,329
Others	6,276	5,512	5,931	6,241	7,162
Total	**7,23,632**	**7,96,913**	**8,59,734**	**9,10,728**	**9,69,680**

Source: Transport Commissioners Office Maharashtra, Mumbai, 2000

Specific problem of Indian cities is the plying of ill maintained public and transport vehicles. Urban buses in India have frequent start and stops, carry more passengers than the designed capacity, hence unable to attain top speed.

Average service life and annual mileages of Indian Vehicles is as given below-

Vehicle Type	Service Life Year	Annual Mileage in Kms
2-Wheeler	15	10,000
3-Wheeler	10	40,000
Passenger car	20	15,000
Taxis	10	30,000
MUV	15	37,000
Trucks	15	40,000
Buses	8	30,000
LCV	5	60,000

MUV-Multi-utility Vehicles; LCV-Light Commercial Vehicles
Source: Transport Fuel Quality for Year 2005, page 208, Central Pollution Control Board, New Delhi

Fuel

Combustion of fuel by vehicles and industries are the important source of urban air pollution. Type of pollutant depends on type of fuel used. Common petroleum fuels used in India are gasoline and diesel.

In India, petroleum fuel is mainly used for vehicular transport in India. Almost half of the total petroleum fuels in the form of high-speed diesel (HSD) and gasoline are used by motor vehicles. The percentage of energy consumption by vehicular transport in total commercial energy consumption increased from nearly 24% to 31% during 1984 to 1996 [TERI Energy Data Directory & Yearbook (TEDDY) 1998/99] due to sharp increase in vehicle population.

Gasoline

In urban centers transportation sector mainly cause air pollution problems, which in turn widely uses gasoline as fuel. Gasoline is used in private vehicles, hence is the biggest contributor to *transportation energy* use in automobile dominant countries like USA and Australia. Gasoline is composed of a blend of light distillate *hydrocarbons*, including *paraffins*, olefins, *naphthenes* and aromatics. It has hydrogen to carbon ratio varying from 1.6 to 2.4. A typical formula to characterize gasoline is C_8H_{15}, with a molecular weight of 111. A high hydrogen content gasoline is C_7H_{17}. Gasoline in USA has low benzene (1.6% by volume) and sulphur (0.038% by weight) content in comparison to India (3-5% by volume and 0.1% by weight respectively). *Lead* has been added to gasoline in the form of *tetra alkyl lead* to raise octane levels and help engine run more smoothly. It is emitted as tiny particles in exhaust contaminating air, soil and vegetation. Emission from gasoline combustion mainly includes hydrocarbons, oxides of sulphur and lead.

*Diesel**

Diesel is mainly used in commercial vehicles but nowadays its use in private vehicles is also increasing. In India, the ratio of diesel usage to gasoline usage is about 7:1. In most other countries, this ratio is about 1:1. Diesel fuels, like gasoline are mixtures of paraffin, olefin, naphthalene, and aromatic hydrocarbons, but their relative proportions are different.

---

* *Source*: Bureau of Indian Standard and www.osc.edu/research/pcrm/emissions/petrol.html.

Diesel fuels have about 8% greater energy density by volume than gasoline and are much less flammable. Emissions from diesel combustion include HC, NO_X, sulfates, soot, and Particulate Matter (PM). The particulates are formed mainly during the diffusion-burning period in Compression Ignition (CI) engines. The flame propagates towards the core of the fuel spray having larger fuel droplets than the periphery, which coupled with richer local air-fuel ratio at the spray core results in incomplete combustion and formation of carbonaceous particles (Horrocks 1994). Diesel in India has higher carbon content ($C_{16}H_{34}$) and molecular weight (226.43 and 198.04 respectively) in comparison to US ($C_{14.4}H_{24.4}$).

*Natural Gas**

As cities are becoming more transport oriented and pollution problems due to gasoline and diesel are visible, natural gas is emerging as an alternative fuel. Natural gas is primarily composed of methane, about 90 to 95%, with small amounts of (0-4%) nitrogen, 4% ethane, and 1 to 2% propane. Methane has higher hydrogen to carbon ratio relative to gasoline and therefore CO_2 emissions from natural gas burning are about 22 to 25% lower than gasoline. It is used in a compressed (CNG) form for vehicular transport. Natural gas has an octane number (RON) of about 127, and therefore natural gas engines operate at a compression ratio of 11:1, greater than gasoline fueled engines (compression ratio of 8:1 and RON 91-99). Sulphur content in CNG of India (Kandla) is higher (100 ppm) in comparison to USA (Texas) (International Petroleum Specification, Inc.1993).

Natural gas is pressurized to in vehicular storage tanks, so that it has about one-third of the volumetric energy density of gasoline. Natural gas does not require mixture enrichment for cold starting, reducing the cold start HC and CO emissions. The combustion of methane involves only carbon-hydrogen bonds, and no carbon-carbon bonds, and hence combustion is more complete, producing less non-methane hydrocarbons. The particulate emissions of natural gas are very low relative to diesel fuel. Natural gas has a lower adiabatic (approx. 224^0 K) than gasoline (approx. 231^0 K). The lower combustion temperatures lower the NO formation rate, and produce less engine-out NO_x. Carbon monooxide

* *Sources:* www.osc.edu/research/pcrm/emissions/petrol.html.

and NO_x emission from CNG fuelled engines (CO- 0.69, 0.01; NO_x- 0.015 and 0.10 g/mile for 2.2 L and 2.2L bi fuel engine respectively) are small in comparison to gasoline (CO- 1.54, NOx- 0.17 g/mile). The emissions of toxics (Benzene, 1,3 Butadiene, Formaldehyde and Acetaldehyde) were highest in gasoline-operated engines (40.6mg/mile), then in CNG start and gasoline run (19.3mg/mile) and least in CNG fuelled engines (3.8mg/mile).*

Fuel Combustion

Combustion of fuel is an important factor because it determines the quantity and quality of pollutants released. Incomplete combustion of fuel releases larger quantity of pollutants in comparison to complete combustion. Combustion of fuel depends on the type of engine. There are two types of reciprocating internal combustion engines used for vehicular transport, gasoline or spark ignition (SI) engines and diesel or compression ignition (CI) engines. Fuel combustion depends on air and fuel mixture ratio. In an SI engine, petrol and air in relatively constant proportions are mixed in the carburetor. This mixture is drawn into the engine, compressed and is ignited by a spark. For maximum power, idling, acceleration, and deceleration fuel-air ratio is richer than ideal (stoichiometric). Incomplete combustion implies unburned hydrocarbons, partially burnt (CO), aldehydes *etc*., in the exhaust. CO is mainly emitted from SI engines and particulates are released mainly from CI engine.

***Engines* for vehicular transport**

There are two types of reciprocating internal combustion engines used for vehicular transport, gasoline or *spark ignition (SI) engines* and diesel or *compression ignition (CI) engines.*

In a SI engine, petrol and air in relatively constant proportions are mixed in the carburetor. This mixture is drawn into the engine, compressed and is ignited by a spark. For maximum power, *idling*, acceleration, and deceleration fuel-air ratio is richer than ideal (stoichiometric). Hence, combustion is incomplete except when cruising. Incomplete combustion implies unburned hydrocarbons, partially burnt

---

* *Sources :* www.osc.edu//research/pcrm/emissions/petrol.html.

(CO), aldehydes *etc*., in the exhaust. Most extreme conditions are encountered when starting up from the cold. Old engines also issue as a bluish smoke the lubricating oil, which has leaked into the combustion space. Some time a vehicle is driven with one or two cylinders not firing at all. In such case one sixth, one quarter or even one-third fuel is pumped into the atmosphere in an unburned condition. This also happens when misfiring occurs. SI engines use gasoline or CNG as fuel.

In a CI engine, a low volatility fuel is converted from a cold liquid into a finely atomized state, vaporized, and its temperature is raised to support auto ignition. The time interval between the start of ignition and combustion is termed the ignition delay. Once region of vapor-air mixture as it is first injected into the cylinder is at or above auto ignition temperature, it will spontaneously ignite. This phenomenon is known as premixed combustion. Finally, the fuel in the main body of the fuel jet will mix with the surrounding air and burn, which is called mixing controlled combustion phase. Combustion occurs at a rate at which the remaining fuel to be burned can be mixed with air. In compression engines, only air is compressed instead of air-fuel mixture and the fuel, directly injected into cylinders, ignites without a spark by contact with the heat from the compressed air. Fuel-air ratio is lean than ideal (stoichiometric) and hence CO and hydrocarbon emissions are minimal. Hydrocarbons from diesel engines are emitted primarily from (i) fuel trapped in the injector at the end of injection that later diffuses out, (ii) fuel mixed into air surrounding the burning spray so lean that it can not burn, and (iii) fuel trapped along the walls by crevices, deposits, or oil due to impingement by spray. CI engines use diesel as the fuel. Table given below gives a comparison of SI and CI engines.

SI and CI engines could be four-stroke (four-cycle) or two-stroke (two cycle). Two-stroke engines are widely used in India to power mopeds, scooters, motorbikes, and small three-wheeled vehicles, due to low initial and maintenance costs. Four-stroke engines have superior fuel efficiency as against the superior (theoretically double) power output from the two-stroke engines for the same speed. The important differences between four-stroke and two-stroke engines result primarily from the scavenging process and the different exhaust gas stream it produces. There are two types of two-stroke engines. One where the fuel is mixed with scavenging air prior to entry into cylinder, the other

Comparison of SI and CI Engines

SI Engine	**CI Engine**
Constant volume cycle	Constant pressure cycle
Premixed flame	Diffusion flame
Homogeneous fuel-air mixture	Heterogeneous fuel-air mixture
Fuel rich combustion Spark ignited	Lean fuel or stoichiometric combustion Self-ignited
High speed engines Unburned fuel and CO emissions	Low speed and high power engines Unburned fuel and CO emissions are minimal.
Gasoline is more volatile and has 0.1% or less sulfur	Diesel is less volatile and has 0.3 % sulfur
More frequent incomplete combustion	Less frequent incomplete combustion
Predominant source of CO	CO emissions only during smoke, major source of particulate matter and aldehydes emissions

where the fuel is injected directly into the engine cylinder after the scavenging process. Three simple scavenging models are shortcircuiting, perfect mixing, & perfect scavenging. In perfect scavenging, no mixing occurs and air simply displaces the exhaust gas to expel it. In the case of short-circuiting, the air initially displaces all the gas (exhaust) within the path of short circuit and then simply flows into and out of cylinder along that path. In simple ported two-stroke engines, a carbureted crankcase-scavenged engine typically loses 15-20% of its fuel due to short-circuiting. For perfect mixing, the first air to come in is instantaneously mixed with the exhaust and the first gas expelled is low in purity, being nearly all residual gas. The exhaust from a two-stroke engine contains both unreacted fuel that did not burn or react at all, as well as organic compounds formed from partial reaction of the fuel molecules. Oil added to the fuel or air stream to lubricate a piston in a two-stroke engine also may not fully burn. The emissions from two-stroke engines include NO_x, CO, CO_2, HC, aldehydes, particulates, and SO_x.

Exhaust emissions

Vehicular transport using Internal Combustion (IC) engines has a major share in the emission of air pollutants such as, particulate matter, lead, Hydrocarbons (HC), Carbon monoxide (CO), Sulfur dioxide (SO_x), Nitrogen oxide (NO_x), non-methane organic compounds and greenhouse gases, particularly Carbon dioxide (CO_2). Combustion of petroleum fuels in IC engines is considered to be one of the dominant *anthropogenic* sources of these polluting gases. Emissions from these vehicles account for roughly half of NO_x, CO, and HC pollutants in the air. Diesel engines also emit irritating and malodorous fumes and smoke. In India, there has been a rapid growth of the vehicles, especially two and four-wheel vehicles since 1990's, and as a result in the level of polluting gases.

Quantities of pollutants (*Giga gram*) emitted from transport sector in India for the years 1990 and 1994 are given in the table below. The emission estimates for 1990 and 1994 in India based on the mass emission factors of CO, HC and NO_x for Indian driving conditions and IPCC emission factors are given prepared by the Central Road Research Institute, New Delhi.

Quantities of pollutants (*Giga gram*) emitted from transport sector in India for the years 1990 and 1994

Species	Gasoline				Diesel			
	MC/SC		Cars		LCV		HCV	
	1990	1994	1990	1994	1990	1994	1990	1994
CO_2	7020	7670	3780	4130	13950	17400	32550	40600
NO_x	5.2	5.6	24.2	26.5	456.4	569	2198	2741
CO	1352	1470	214.2	234	217.2	271	1710	2132
NMVOC	670	728	42	44	231.8	289	427.4	533

Source: Reference 3 NMVOC = Non Methane Volatile Organic Compounds, MC = Motor Cycle, SC = Scooter, LCV = Light Commercial Vehicles, HCV = Heavy Commercial Vehicles.

These are aggregate emission estimates based on default emission factors. Modeling based on fuel characteristics and combustion technology is needed to calculate the detailed emission factors.

Emissions from internal combustion engines

The exhaust of an IC engine contains one or more of the following emissions.

1. *Carbon dioxide (CO_2)*
2. *Carbon monoxide (CO)*
3. Unburned *hydrocarbons (HC)*
4. *Oxides of nitrogen (NO_x)*
5. *Water* vapor (H_2O)
6. *Oxides of sulphur (SO_2)*
7. Oxygenated hydrocarbons like *aldehydes*
8. *Smoke*, soot, and other forms of black carbon
9. *Particulate matter* including lead

While CO_2 emissions are of major concern for climate change, CO, NO_x, HC, and particulate emissions also affect the human, animal, and plant health.

Four recognized sources for emission of polluting gases in vehicles with IC engines are:

(a) The exhaust pipe
(b) The crank case
(c) The carburetor
(d) The Fuel Tank

The design and condition of engine, operating conditions, and ambient air characteristics have considerable influence on the nature and type of emissions. Typical values of emissions of CO, HC and NO_x in parts per million under different operating conditions are given in table:

Typical values of emissions (in ppmv) under different operating conditions

Constituents	Idling	Cruising	Acceleration	Decelerating
CO	6400	24000	24000	45000
Hydrocarbons	1400	620	810	5700
NO	0	1400	1700	0

The common constituents of hydrocarbons in automobile exhaust are:

Paraffins	C_nH_{2n+2}	Methane, Ethane, Propane, Isobutane
Olefins	C_nH_{2n}	Hexene, Butadiene
Aromatics	C_nH_{2n-6}	Benzene, Toluene
Acetylens	C_2H_2	
Naphthene	C_nH_{2n}	Cyclopentane

Typical exhaust gas compositions from the sampling of a very large number of four wheelers (passenger cars) undergoing an arbitrary typical diving cycle in USA are given in below

Typical Exhaust Gas Composition

Mode of operation	Unburned hydro-carbons (PPM)	CO (Vol.%)	NO_x (PPM)	Hydrogen (Vol.%)	CO_2 (Vol.%)	Water (Vol.%)
Idle	750	5.2	30	1.7	9.5	13.0
Cruise	300	0.8	1500	0.2	12.5	13.1
Acceleration	400	5.2	3000	1.2	10.2	13.1
Deceleration	4000	4.2	60	1.7	9.5	13.0

Source: Urban Air Pollution in Mega cities of the World: Published on behalf of the World Health Organization and the United Nations Environment Programme by Blackwell, 1992.

Modeling of the emissions is based on the chemical kinetics. Emissions depend on fuel composition, temperature, pressure, and the availability of oxygen for reaction to occur.

Carbon monoxide

Carbon monoxide appears in the exhaust of rich-running engines. One can compute the carbon monoxide concentration using equilibrium assumptions. SI engines usually run on lean fuel except during idling and hence CO is not a problem in other stages of driving except idling.

Nitrogen oxides

Nitrogen oxides (NO_x) are formed during the combustion process due to the reaction of atomic oxygen and nitrogen. NO_x emissions are relatively low during engine start and warm-up as the reactions forming NO_x are temperature dependent. NO concentrations are maximum for slightly lean mixtures. Dilution of the charge by residual gas, explicitly via exhaust gas re-circulation, implicitly via throttling, or by moisture in the inlet air reduces the NO.

Unburnt hydrocarbons (HC)

A large fraction of the fuel escapes burning during normal combustion in engines. In a four-stroke SI engine, approximately 7 - 9% and in a two-stroke SI engine more than 20% of the fuel supplied to the engine is not burnt during the normal combustion phase. Some of the fuel will oxidize within the cylinder as it mixes with the burned gases, some may oxidize in the exhaust system before the exhaust gas becomes too cool for hydrocarbon burn up to occur. Thus the engine-out HC emissions are less than the amount that escapes burning during the normal combustion process. Other sources of HC emissions are evaporative losses from the fuel system, the crankcase blow-by gases in a two-stroke engine, and a small fraction of the in-cylinder gases blow by the piston rings into the crankcase. Important sources of HC emissions in a two-stroke engine and a four stroke engine are significantly different.

The distribution of pollutants according to source within vehicle is:

(a) Tank and carburetor (evaporation losses): 20% of the hydrocarbons

(b) Crank case blow by: 20% of the hydrocarbons

(c) Exhaust: 60% of the hydrocarbons and almost all of the remaining listed above

The relative amounts of various hydrocarbon compounds from a carbureted crankcase-scavenged two-stroke SI engine and a similar-capacity four-stroke engine are shown in the following table

Comparison between gas chromatographic analysis of HC emissions

HC Component (% weight)	Formula	Two-stroke	Four-stroke
Methane	CH_4	1.00	4.21
Acetylene	C_2H_2	0.99	7.81
Isopentane	C_5H_{12}	8.42	5.38
2-Methyl pentane	C_7H_{16}	4.13	2.98
3-Methyl pentane	C_8H_{18}	2.67	2.06
n-Hexane	C_6H_{14}	2.85	2.38
3-Methyl hexane	C_9H_{20}	1.71	1.31

Table given below gives the relative hydrocarbon emissions from different sources in a SI Engine.

Hydrocarbon emission sources in a four-stroke SI engine.

Source	% Fuel escaping normal combustion	% HC emissions
Crevices	5.2	38
Oil layers	1.0	16
Carbon deposits	1.0	16
Liquid fuel	1.2	20
Flame quench	0.5	5
Exhaust valve leakage	0.1	5
Total	**9.0**	**100**

The crevice mechanism is the most significant, responsible for about 40% of the hydrocarbon emissions. In a CI engine, hydrocarbon emissions come primarily from; (1) fuel trapped in the injector, (2) fuel mixed into air surrounding the burning spray so lean that it can not burn, and (3) fuel trapped along the walls by crevices, deposits, or oil due to impingement by spray.

Smoke, soot, and other forms of black carbon

A high concentration of Particulate Matter (PM) is manifested as visible smoke or soot in the exhaust gases. The diffusion flame in CI engines causes the formation of larger soot particles at the jet periphery. Most of the soot is burned with the fuel at the diffusion flame. The fraction of soot that is not oxidized becomes an exhaust emission. Some of the in-use four-stroke engines, with highly worn out piston rings or valve guides, also lead to the emission of visible smoke due to an increased consumption of lubricating oil that fails to burn. Smoke forms in diesel engines due to heterogeneous combustion. Fuel rich combustion occurs both in the premixed and the mixing controlled phases of combustion.

The organic fraction results from all the processes that generate hydrocarbons and their partial oxidation products. During the dilution process some of them cool enough to condense or adsorb the soot.

The average composition of particulate matter reported from 16 engines are shown as (Needham *et al*, 1989):

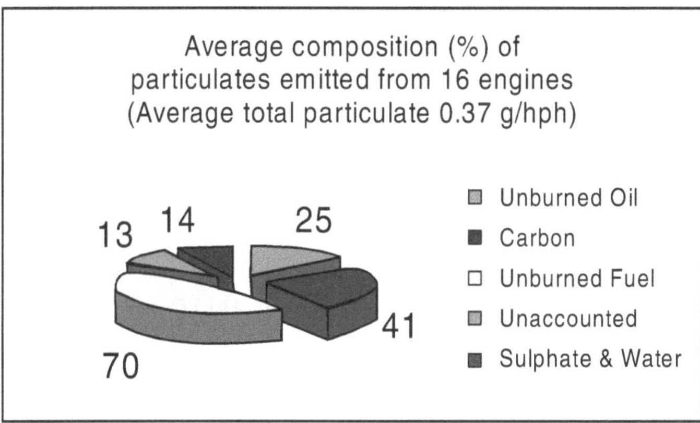

Fuel consumption

Despite rules and regulations for vehicle emission urban air quality is continuously deteriorating, reason may be due to continuous rise in vehicle population and fuel consumption. In India, petroleum fuel is mainly used for vehicular transport. Almost half of the total petroleum fuels in the form of high-speed diesel (HSD) and gasoline are used by motor vehicles. The percentage of energy consumption by vehicular transport in total commercial energy consumption increased from nearly 24% to 31% during 1984 to 1996 [TERI Energy Data Directory & Yearbook (TEDDY) 1998/99] due to sharp increase in vehicle population. As a result of large concentrations of vehicles, the mega-cities of India are major consumption points of petroleum fuel. For example, the total petrol and Diesel supplied to the Mumbai by BPC, IOC, HPC and IBP during 1998-99 was 372197 KL, 416362 KL and in 1999-2000 it rose up-to 363841 Kl and 410109 Kl respectively. In Mumbai, the fuel consumption in terms of petrol and high-speed diesel is approximately 30320 and 34175 Kl per month respectively.

Petroleum fuel consumption in Mumbai City

Agency	**Petrol**		**Diesel**	
	1998-99	**1999-2000**	**1998-99**	**1999-2000**
BPC	1,61,492	1,54,029	1,90,939	1,83,153
IOC	70,203	67,579	77,788	74,695
HPC	1,31,067	1,30,861	1,28,921	1,37,799
IBP	12,435	11,372	18,714	14,462
Total	3,72,197	3,63,841	4,16,362	4,10,109

All figures are in kilo liters (kl)

In Delhi, the consumption of petroleum fuels also increased by more than three times in 1996-97 compared to 1980-81 as reflected in table:

Petroleum fuels consumption in Delhi ('000 tons)

Item	**1980-81**	**1990-91**	**1992-93**	**1993-94**	**1994-95**	**1995-96**	**1996-97** (Provisional)
Gasoline	133	344	363	375	408	436	449
High Speed Diesel Oil	377	732	810	840	929	1153	1234
Light Diesel Oil	59	66	41	67	64	45	49

Chapter - 3

MAJOR AIR POLLUTANTS IN URBAN AIR

Major air pollutants in urban air and their sources are:

SO_2 : Burning of fuel, Industrial Processes

NO_x : Burning of gasoline, natural gas, coal, oil *etc*, Vehicle exhaust, Power Stations

SPM : Burning of fuel, Vehicle exhaust, Industrial sources, burning of Refuge, re-suspension of dust

CO : Burning of gasoline, natural gas, coal, oil *etc*.

Lead : Leaded gasoline, paint, smelters, Manufacturing of Lead storage batteries

Ozone : Chemical reaction of pollutants, VOCs and NO_x in the presence of heat and sunlight.

VOCs : Released from burning fuel (including Natural Gas), solvents, paints, glues. Cars are important source of VOCs

Major sources of air pollution in urban areas are:

AIR POLLUTION FROM *AUTO EXHAUST*

Motor vehicles are the major source of a number of pollutants, in particular carbon monoxide (CO),nitrogen oxides (NO_x), unburnt hydrocarbons (HC),ozone (O_3) and other photo chemical oxidants and lead (Pb) and, in smaller proportions Total Suspended Particulate Matter (TSPM), sulphur dioxide (SO_2) and volatile organic compounds. These pollutants result either from primary emissions or atmospheric transformations. Comprehensive data are not available on air pollution emissions from transportation *etc*. However, available data exhibit that motor vehicles are the source of emissions of CO, NO_x and volatile organic compounds (HEI,1988). In urban areas where pollution levels tend to be highest, the motor vehicles contribution tends to be even higher. Usually, over 90% of the CO in the city centers comes from vehicles and its common to find 50 to 60% of the HCHO and NO_x coming from this source (OECD, 1987).

Air Pollution From Industries

A large number of industries such as chemical industries, rubber, textile, iron and steel, non-metal products, food products, paper, printing, metal products and leather industries are responsible for about 20% of air pollution. The Central Pollution Control Board has identified 1551 medium and large industrial units in the country in 17 highly polluting industrial sectors. Out of these about 15% are air polluting and 8% are both air and water polluting, The industrial sector in 1995 contributed 2 million metric tonnes of pollutants (IGIDR, Mumbai).

Pollutants released from different industries are CO_2, CO, SO_2, H_2S, NO, NO_2, smoke, particulate matter organic and inorganic gases. Some major pollutants and their industrial sources are given below

Pollutant	**Industry**
Sulphur dioxide	Thermal power plant, coal and oil combustion
Oxides of Nitrogen	Industries using coal as fuel
Hydrogen sulphide	Refineries, chemical industries
Carbon mono oxide	Industries using coal as fuel
Hydrogen cyanide	Chemical industries, metal plating *etc.*
Ammonia	Dye and fertilizer industries
Phosgene and carbonyl chloride	Chemical and dye industry
Aldehydes	Oil industry
Arsines	Metal and acid industry
Suspended particles	Almost every manufacturing process.

Petroleum refineries emit SO_2 and NO_x into the atmosphere. Cement industry emit SPM five times greater than industrial safety limits. Thermal power plants are the principal source of SO_2 in air.

A variety of industrial units in Delhi have increased from 20,000 to 55,000 and about 44,000 of these are located in residential areas and emitting SO_2, NO_x, Particulate matter and acid vapors in air. Major pollutants emitted by the Delhi industries are given below (Sharma and Kaur, 97):

Area	Number of Air polluting units	Pollutants (Tonnes)	
		SO_2	Dust
Lawrence road	28	20.5	140.2
Wazirpur	90	18.2	254.5
Kirti Nagar	14	3.0	66.5
Moti Nagar	3	1.1	33.4
DLF Area	15	2.1	56.2
Nazafgarh road area	32	75.3	7952

Air Pollution From Bio-fuels

Burning of bio fuels is one of the sources of air pollution in cities. Burning of unprocessed cooking fuels in homes mainly accounts for air emission in India IGIDR, Mumbai. About 75% of Indian households still relay on bio-fuels such as cow dung, fuel wood and crop residue. Some 82% of sulphur dioxide emissions and 39% of nitrogen di oxide emissions are produced within the home. Some 96% of the particulate matter emissions in the country also come from the household sector. In Indian cities population lives in different localities like independent houses, apartments, slums *etc*., Population living in slums uses low quality bio fuels.

Other Sources

Burning of solid waste also adds to the air pollution. Uncontrolled burning of solid waste in urban and semi urban areas increases air pollution level in cities. Mumbai city with the resident more than 10 million people produces 5,000 tonnes of garbage every day. An IGIDR study estimates that burning one ton of garbage produces 0.988 tonnes of suspended particulate matter, 0.072 tonnes of sulphur dioxide, 0.136 tonnes of volatile organic chemicals, 0.18 tonnes of nitrogen dioxide and 0.38 tonnes of carbon monooxide. In the study it was estimated that 307 million inhabitants of 3,696 cities and towns in India would release 1.18 million tonnes of pollutants annually. If the emissions produced by burning waste in rural areas were also included, total emissions by waste burning would be equivalent to more than two million metric tonnes emitted by the industrial sector in 1995.

Major Air Pollutants

Suspended particulate matter (SPM)

It has quite recently been realized that particulate pollution is a very serious problem in urban areas. SPM consist of solid and liquid particles emitted from numerous natural and man-made sources and a complex and variable mixture of different sized particles with many chemical components. Larger particles may come from wind blown dust, volcanic eruptions, pollen/ spores and algal filaments, finer particles tend to be formed by combustion and gas to particle conversions (chiefly from SO_2). The constituents may vary, although in urban areas they typically include carbon particles and poly-nuclear aromatic hydrocarbons produced by incomplete fuel combustion. The modern day SPM problems are aggravated by the lack of emission regulation on the increasing number of diesel powered vehicles used for transport. Exhaust from diesel powered vehicles is a major source of SPM (Beckett *et. al.*, 1998).

Particulate matter, or PM, is the term for particles found in the air, including dust, dirt, soot, smoke, and liquid droplets. Particles can be suspended in the air for long periods of time. Some particles are large or dark enough to be seen as soot or smoke. Others are so small that individually they can only be detected with an electron microscope. Some particles are directly emitted into the air. In urban areas particles come from a variety of sources such as cars, trucks, buses, factories, construction sites, tilled fields, unpaved roads, stone crushing, and burning of wood (Street *et. al.*, 1996).

Convention splits the definition of 'health-risk' air- borne particles into a number of size classes (QUARG, 1996) of which the coarse and fine fractions are the simplest. The coarse fraction includes all suspended particles in the PM_{10} size range above 2.5 μm in aerodynamic diameter, while the fine fraction (PM2.5) includes all those remaining. Arbitrarily, the course fraction is made up of natural and organic particles, while the fine fraction contains mostly particles of anthropogenic source, including soot, nitrate and sulphate particles (DOE, 1995a). A common measure of particles used to quantify pollution concentrations and their effects is the PM_{10} value. PM_{10} is the abbreviation for particulate matter having an aerodynamic diameter of less than 10 μm and is measured in μgm^{-3} of air. This is the size range, which includes the majority of the total suspended mass of particles in the atmosphere.

Motor vehicle emissions usually constitute the most significant source of ultra fine (>$PM_{2.5}$) particles in an urban environment (Zhu *et. al.*, 2002b, Technical Paper). Ultra fine particles are of high concern because they cause health problems as well as affect the environment. Zhu *et.al.* (2002a) conducted systematic measurements of the concentration and size distribution of ultra fine particles in the vicinity of a highway dominated by gasoline vehicles. The particle number concentration was found in the range of $1.8x10^5$ to $3.5x10^5/cm^3$ at 17 km from the highway in down wind direction. Atmospheric dispersion and coagulation appears to contribute to the rapid decrease in particle number concentration and change in particle size distribution with increasing distance from the freeway. Average traffic flow during the sampling period was 12180 vehicles/hr. with more than 25% of vehicles being heavy duty diesel trucks.

Aerosols always remain in air, has a great importance because of their hazardous effect. These are formed from the

- » Condensation and sublimation of vapor.
- » Production of smoke.
- » Chemical reactions in the atmosphere involving trace gases.
- » Mechanical disruption of the earth's surface, forming dusts over landmasses and salt crystals over the oceans.
- » Coagulation of small particles to form larger ones (Cadle, 1996).

The UK national inventory of PM_{10} from anthropogenic primary sources (1993) reproduced from DOE (1995a), highlight the importance of road-transport emissions, that total emission was 263 thousand tonnes (Fossil fuel 40, Domestic 37, Commercial 5 Refineries and Industrial Combustion 45, Construction 4, Industrial Process 30, Mining and Quarrying 29, Road Transport 66, Other Transport 8000 tonnes).

Sulphur dioxide (SO_2)

Sulphur dioxide (SO_2) is an important gaseous pollutant, arising from both natural and anthropogenic sources in urban air. Natural sources such as sulphur bacterial activities, volcanoes, forest fires *etc.*

contribute to environmental levels of SO_2. Man made contributions include the use of sulphur containing fossil fuels for transportation, domestic purposes and power generation. Sulphur oxides (SO_x) are compounds of sulphur and oxygen molecules. Sulphur dioxide is the predominant form found in the lower atmosphere. It can be detected by taste and smell in the range of 1000-3000 $\mu g/m^3$. At concentrations of 10,000 $\mu g/m^3$ it has a pungent unpleasant odor. SO_2 dissolves readily in water present in the atmosphere to form Sulfurous acid (H_2SO_3). About 30% of the SO_2 in the air is converted to sulphate aerosol, which is removed through wet or dry deposition processes.

Background level of SO_2 in the atmosphere is 0.001 $\mu g/L$ or less (Kellogg *et. al.,* 1972). About 10^9 million metric tons of SO_2 are estimated to be added to the earth's atmosphere every year. As more countries industrialize, acidification of environment is becoming a global problem (Helbwachs, 1983). Asian emissions of SO_2 were expected to exceed those of Europe and North America combined in the year 2000.

Man made SO_2 emission has been rising about 4% annually, a rate equivalent to the rise in global energy consumption. About 90% of global SO_2 emissions come from the Northern hemisphere. The developing countries will contribute more as they develop their industrial base. SO_2 pollution is becoming particularly evident in countries such as China, Mexico, and India. The dispersal area, the amount of acid produced and the polluting effects of SO_2 depend on factors such as weather conditions (wind, cloud cover, humidity and sunlight), the presence of other pollutants, the height at which SO_2 is released and the length of time it stays in atmosphere. SO_2 emission is increasing in developing countries like India, may be due to the continuous increase in vehicle population and lack of effective norms and regulations (Yunus *et.al.*, 1996).

Oxides of Nitrogen (NO_X)

Oxides of nitrogen are released in all the types of combustion as they are formed by the oxidation of atmospheric nitrogen at high temperature. Nitric oxide usually emitted from the automobile exhaust is oxidized to nitrogen dioxide (NO_2) by reaction with oxidants (prominently ozone) present in the ambient air. Nitrogen dioxide is a reddish brown gas with a characteristic pungent odor. It is corrosive and a strong oxidizing agent. Nitrogen dioxide is the predecessor of gaseous nitric acid and nitrate aerosols, which has the biggest health impact. The major sources of NO_2 are combustion-associated processes, such as motor vehicles, power plants as well as any high temperature combustion process used in industrial work. Oxides of nitrogen particularly nitrogen dioxide are toxic gases. Chemical properties of NO_x are:

Substance name- Oxides of Nitrogen

CASR number- NO_2 - 10102-44-0
NO - 10102-43-9
N_2O - 10024-92-2

Molecular Formula NO_2
NO
N_2O

Synonyms-
NO_2: nitrogen dioxide
NO : nitric oxide, nitrogen monooxide, mononitrogen monoxide N_2O: nitrous oxide, dinitrogen monoxide, laughing gas, and hyponitrous acid anhydride

Physical and chemical properties

NO_2 is a dark brown, fuming liquid or gas with pungent smell, acidic odour, and detectable at 0.12 ppm. NO is a colorless gas with a sharp, sweet odour at high concentrations in air. N_2O is a colorless gas with a slight sweetish odour.

Properties of NO_x

	NO_2	**NO**	**N_2O ***
Melting point	-11^{0}C	-164^{0}C	
Boiling point	21.2^{0}C	-151.7^{0}C	-88.6^{0}C
Vapour density	1.58	1.04	1.53
Solubility in water	Highly soluble forms nitric acid (and a strong acid)	Slightly soluble forms nitrous acid (and a weak acid)	

* N_2O is non-inflammable and has anesthetic properties.

The oxides of nitrogen (nitric oxide and nitrogen dioxide) are formed by the direct combination of oxygen and nitrogen during a variety of thermal processes. These include operation of internal combustion engines, thermal power generating plants, and in the air surrounding the arc or flame as a result of heat from electric arc or gas torches (oxidizing flames). In outdoor or open shop welding, hazardous abnormal concentrations are unlikely, except perhaps for short periods. In confined spaces, hazardous concentrations of nitrogen oxides may rapidly build up in welding operations. High concentrations of nitrogen oxides have been found during gas tungsten-arc cutting of stainless steel. Natural sources of oxides of nitrogen are the biological cycling of nitrogen and include processes that produce NO and N_2O as intermediates. Thermal processes in the atmosphere (during lightning) produce oxides of nitrogen from nitrogen gas. While the mobile sources are vehicle exhausts.

Lightning of forest fires and microbial activity in soil also causes natural emissions of NO_x. A 1980 estimates puts a total natural and man made emissions of NO_x at 150 million tonnes per year, over half of it is from natural sources. (UNEP/GEMS, 1991). In the industrial regions of Europe and North America however the man-made NO_x outweighs the natural NO_x by 5 to 10 times.

Carbon monooxide

Carbon monooxide is the most serious air pollutant in the air. Some of the major sources of CO in air are given below

1. It is released in the combustion of fuel:
   $2C+O_2 \rightarrow 2CO$
2. It is also released from blast furnace:
   $CO_2+C \rightarrow 2CO$
3. Carbon dioxide dissociates to form CO
   $CO_2 \rightarrow CO+O$
4. Decomposition of organic matter also yields CO
   $CH_4+OH \rightarrow CH_3+H_2O$
   $CH_3+OH \rightarrow CH_2+H_2O$
   $CH_2+OH \rightarrow CH+H_2O$
   $CH+OH \rightarrow CO+H_2$
5. Decomposition of chlorophyll also gives CO

CO alone forms 48% of all gaseous pollutants. It is produced in atmosphere by natural processes *i.e.* forest fires, natural gas emission, marsh gas production and volcanic eruptions. Anthropogenic activities, mainly vehicular exhaust contributes significantly to CO emission (described in chapter 2).

Hydrocarbons

Trees emit huge quantities of hydrocarbons in air. Plants mostly emit terepenes. Methane is the major naturally occurring hydrocarbon emitted into the air. It is produced during decomposition of organic matter by microorganisms. About 45% of reactive hydrocarbons constitute the source as gasoline while 15%is produced by other activities (described in chapter 2).

In cities automobiles are the chief source of hydrocarbons. Major hydrocarbons in ambient air are methane, ethylene, acetylene, butane, isopentane, toluene *etc*. Hydrocarbons in the presence of NO_x are mainly responsible for the formation of photochemical smog. Photochemical smog is composed of different compounds like O_3, NO_2, H_2O_2, organic peroxide, peroxy acetyl nitrate and CO together with aldehydes and ketonnes.

Ozone

Ozone (O_3) is exceedingly rare in our atmosphere average about 3 molecules of ozone for every ten million-air molecules. Ozone is mainly found in two regions of the earth's atmosphere. Most ozone (about 90%) resides in the layer between approximately 10 and 50 kilometers (about 6 to 30 miles) above the earth's surface in the region of the atmosphere called the stratosphere. The stratosphere ozone is commonly known as the ozone layer .The remaining ozone is in the lower region of the atmosphere called "troposphere" which extends from the earth surface up to 10 kilometer.

The ozone in these two regions, stratosphere and troposphere is chemically identical (both consist of three oxygen atoms and have the chemical formula O_3). The ozone molecules have very different effects on humans and other living things depending on their locations. Stratospheric ozone plays a beneficial role by absorbing most of the biologically damaging ultraviolet sunlight called *uv–b* radiations, thus allowing only small amount of ozone to reach the earth surface.

Chemistry of ozone

Physical properties

Molecular formula	:	O_3
Molecular weight	:	48.0
Physical state at NTP	:	Colorless gas (Light blue)
Odor	:	Characteristic smell at 0.02 ppm
Melting point (M.P.)	:	-192.7 + 0.2°C
Boiling point (B.P.)	:	-111.9 + 0.3°C
Vapor density	:	
At 0°C 101 KPa (760mm Hg)	:	2.14 g/lit.
At 25°C101 KPa (760mm Hg)	:	1.96 g/ lit.
Specific gravity relative to air	:	1.658
Solubility at 0°C	:	0.494 ml / 100 water

Ozone is an unstable triatomic allotropic form of oxygen. It is light blue gas, which condenses at –120°C as a dark blue liquid. The liquid freezes at –193°C. It is one of the strongest oxidizing agents. The oxidation potential in acid solution is =2.07V. This is represented as –

$$O_3 + 2H^+ + 2e_ \rightarrow O_2 + H_2 O : E^0 = +2.07 \text{ V}$$

Conversation factors:

At 25°C .760 mm Hg	1ppm =2000 μg/m^2 1 g/m^2=5.097x10^{-4}
At 00C 760mmHg	1ppm=2141 μg/m^2 1gm^3 =4.670 x10^{-4}

In the troposphere (with PBL), the ozone is produced through the photo dissociation of NO_2 (source of atomic oxygen) at >- 420nm. The increase in the ambient ozone level in urban and even in rural areas has been primarily due to the presence of "photochemical smog produced through increased industrial activity. CO and ozone show a positive relationship in areas of higher NO_x where ozone is being created photo chemically.

Chapter - 4

Global Status of Major Air Pollutants

Particulate Matter

Anthropogenic emissions of particulate matter include fossil fuel (diesel and petrol) combustion from mobile and stationary sources (Rogge *et. al.,* 1993a-c), biomass burning (Forest fires, landfills, agricultural activities and incinerators) (Oros and Simoneit, 1999) natural gas (Rogee *et. al.,* 1993c), wood (McDonald *et. al.,* 2000) and coal combustion (Mastral and Callen, 2000).

According to WHO's 1999 guidelines of Air Pollution Control (Air pollution) the concentration of SPM is decreasing in developed countries due to strict rules and increasing in developing countries (Yunus *et. al.,* 1996). In developed countries data for the period 1982-1984 show emission of about 27 million tonnes of SPM per year, but the current total is probably closer to 135 million tonnes (UNEP/GEMS 1991), SPM emission has fallen in most industrialized countries between 1970 and 1990, particularly in recent years due to massive and effective air pollution reduction programmes.

In London enforcement of clean air acts (1956, 1964, and 1968) has resulted in dramatic reduction in SPM. The reduction in solid fuel use, together with introduction of smokeless fuel, caused a steady decline in emission through the 1970s and 1980s. In 1983 total smoke emissions were 11000 tonnes compared to the 14000 tonnes in 1978 (WHO/UNEP 1992). Transport accounts for 76% of total emission of SPM in London. Smoke emission from diesel-engine vehicles was estimated to be 66% of the total smoke emission in 1984. The annual SPM concentration has now come below 30 $\mu g/m^3$.

Particulate emission is increasing in developing countries due to lack of sustainable development and poorly maintained vehicles or absent of abatement equipments. WHO (1999) has recommended guidelines for annual and daily emission of SPM (annual mean of 60-90 μg/m^3 and a daily average of 150-230 μg/m^3). A number of cities in developing countries exceed the guideline on particulate.

Particulates that range in size from less than 0.1 micrometer (μm) upto approximately 45 μm are designated as dust or Total Suspended Particulates. Particulates larger than that range tends to settle as dust and do not remain suspended, except during high winds. Particulates less than 10 microns diameter is called as respirable particulate matter.

Study by Shen *et. al.* (2002) has lead to the creation of new particulate matter standard with diameters < 2.5μm ($PM_{2.5}$). A particle measurement campaign was conducted by Titta *et. al.*, (2002) in suburban environment near a major road in Kuopio, Central Finland in August to September 1999. The mass concentrations of fine particles ($PM_{2.5}$) were measured at distances of 12, 25, 52 and 87 m from the center of a major road at a height of 1.8 m during 6.00 am to 10.00 pm for 27 days. Traffic flow and relevant meteorological parameters were also measured. The regionally and long range transported contribution, the primary and non-exhaust vehicular emissions, and other sources were estimated to contribute on an average 41+ 6, 33+ 6 and 26+ 7% of the observed $PM_{2.5}$ concentrations respectively.

Deposition of particles from the atmosphere occur by the action of wind turbulence on a large scale when wind speed falls in the laminar sub Layer due to the presence of a topographical or meteorological obstruction; *e.g.* a forest/bare ground boundary in the former case and a cold weather front in the latter (Pye, 1987). Larger, heavier particles will fall to earth much quicker than finer particles due to their higher settling velocities. This is the reason for the higher rates of deposition closer to particle emission sources (Boubel *et. al*.., 1994). The residence time of sub-micron particles in the atmosphere is, therefore, much longer than for larger particles. The Quality of Urban Air Review Group's Report (QUARG, 1996) stated that the lower deposition rate results in a 'background' concentration of particles from a variety of widespread sources, covering large areas, and even permeating the indoor environment. The implications of indoor air pollution are relatively new and are discussed by Crump (1996) and Zummo and Karol (1996). In addition, use of plant surfaces to reduce particle concentration in indoor air was discussed by Lohr and Peerson-Mims (1996). Such background levels of particles were measured by Monn *et. al.*. (1995), and found that PM_{10} concentrations at a number of rural areas (Swiss) were very similar due to the mixing of ambient particulates from a variety of disparate sources. They also found that in these areas it was the finer fraction of particles between 0.3 and 0.7 μm diameter that contributed most to the PM_{10} total mass. Janssen *et. al.* (1997) found that the larger diameter component of urban PM_{10} consisted mainly of locally deposited carbon and road dust. These were, therefore, excluded from the finer, homogeneous, background particle mass. This level spread of pollutant is special for particles since gases such as SO_2 and NO_x are removed from the atmosphere relatively quickly due to reaction with other compounds (Score, 1973).

In **New York** total estimated emissions of particulate matter <10μm (PM_{10}) were approximately 55,000-2,90,000 tonnes in 1985 (USEPA, 1991). Over a 25-year period (1965-1990) a striking reduction in annual SPM occurred via enforcement of Federal and New York state particulate emission regulations. The minimization of allowable fuel sulphur content also caused a shift to cleaner, less particulate emitting fuels such as natural gas. SPM decreased slightly from 1975-1989, all well below the upper WHO guidelines of 90 $\mu g/m^3$.

In **Moscow** concentration of SPM for the period 1980-1988 were rather stable with mean annual concentrations of 150-250 $\mu g/m^3$ (MCGNPS, 1991). A considerable decrease was noted between 1988 and 1990 (Rovinsky *et. al.*, 1991). Reduction in the use of liquid and solid fuel by the power stations probably account for the decrease in SPM in last decade. Seasonal variation of SPM is not explicit, but a high increase is observed in April to July and October to December (MCGNPS, 1991).

Air borne particle number concentration and size was monitored at a site within the central business district of Brisbane, Australia by Morawska *et. al.* (2001). Data was correlated with the traffic flow rate on the nearby free way with the aim of investigating differences between weekday and weekend pollutant characteristics. Observations over a five year period shows that mean number particle concentration was higher on weekdays *i.e.* 8.8 x 10^3 cm^{-3} and on weekends 5.9 x 10^3 cm-^{3}, a difference of 47%. The mean traffic flow rate on the freeway was 14.2x10^4 and 9.6x10^4, vehicles per weekday and weekend day respectively. The mean diurnal variations of particle concentration closely followed the traffic flow rate on weekdays and weekends.

In **Tokyo**, SPM levels decreased steadily from 1975 to 1989. Concentrations that were about 55-75 µg/m^3 between 1975-77 came below 60 µg/m^3 at all monitoring sites in 1989 due to strict measurement in all aspect like use of unadulterated fuel, use of alternative fuel and follow the state regulations. High short term pollution levels varied from 130-148 µg/m^3 in the various stations. This is well below the WHO daily guideline of 230 µg/m^3 (Govt. of Japan, 1990).

Levels of Suspended particles PM_{10}, $PM_{2.5}$ and PM_1 were continuously monitored at an urban Kerbside in the metropolitan area of Barcelona by Querol *et. al.* (2001). Hourly levels of $PM_{2.5}$ and PM_1 are consistent with the daily cycle of gaseous pollutants emitted by traffic, whereas TSPM and PM_{10} do not follow the same trend, at least in the diurnal period. The $PM_{2.5}/PM_{10}$ ratio is dependent on the traffic emissions, whereas additional contribution sources for the >10 µm fraction must be taken into account in the diurnal period.

The high concentrations of PM_{10} in California, USA, to their sources was achieved by using a Chemical Mass Balance receptor model, whereby profiles of the elemental composition of particles from suspected sources were compared with ambient particles collected at monitoring stations (Chow *et. al.*, 1996), Results showed that motor vehicles contributed 30-42% of the ambient particle load, road dust 25-27%, and marine aerosol 18-23% in California. These findings illustrate the significant contribution, which anthropogenic sources make to particle concentrations in the urban environment. Such source-apportionment studies reveal the major sources of particles, and are useful for influencing policy and reduction measures aimed at lowering the total ambient load.

Particulate emissions have historically been a significant concern in Lake County (Indiana), and the county continues to have the most serious particulate pollution in the state. In the 1970s and 1980s, ambient levels of total suspended particulates exceeded health standards frequently and by significant margins. In 1993, the Indiana Department of Environmental Management completed a rule making that established new emission limitations for sources in Lake County to meet federal standards for particulate less than ten microns. A strategy implemented by IDEM has sought to control major stationary sources of particulate, including operations at the steel mills. An example of an event, which has reduced particulate, is the closure of the Inland Steel coke batteries. These efforts have resulted in significant emission reductions. The levels of particulate less than ten microns in Lake County have dropped significantly due to new particulate rules and efforts of Lake County industry. The Indiana Department of Environmental Management is beginning to collect air quality monitoring data to assess concentrations of $PM_{2.5}$.

According to the results of program evaluation of air quality in France between 1991-1998, the particles are measured by two manners, on the one hand by the method known as "of the black fume" and on the other hand by method known as "PM_{10}". In 1998, the tendency is with the decrease or stability in the majority of the agglomerations (www.citepa.org).

Air quality in **Dublin** is increasingly affected by emission of pollutants from motor vehicles (Reynolds and Broderick, 2000). A tapered element Oscillating mass was used to measure particulate matter less than 10 microns aerodynamic diameter. The average hourly measured concentration of PM_{10} was 18.31mg/m^3 and the predicted hourly concentration was 13.53 mg/m^3 by an emission inventory.

In **Bangkok** anthropogenic SPM emissions were estimated to be about 40000 tonnes in 1980. Since then emissions have increased by a factor of two (Faiz *et. al.*, 1990). Even in suburban areas SPM concentration exceed the WHO guideline. Particle size analysis of SPM samples have shown that more than 60% of particulates are smaller than 10um and particles in the range of 0.6-1 and 5-7 μm are especially common. This indicates the seriousness of the SPM problem in Bangkok as the particle fall into the inhalable size range (ONEB, 1989).

In **Beijing**, anthropogenic SPM emissions were estimated by World Bank to be about 116000 tonnes in 1985; industry accounts for 59% (68000 tonnes per annum), of it (Krupnick and Sebastian, 1990). Annual mean SPM levels ranged between 200 and 500 $\mu g/m^3$ during 1981-1990, which exceeds the WHO guideline (60-90 $\mu g/m^3$). The mass concentration of TSPM and PM_{10} were determined from four ground based monitoring sites in Xian, China by Zhang *et. al.*, (2002). The samples were collected in four seasons during 1996 -97 and in July 1998. Annual average concentration of TSPM was 410 $\mu g/m^3$. PM_{10} account for 60-70% of TSPM in summer 1998.

In **Karachi** total anthropogenic emission was 106000 tonnes per annum with half of it coming from industrial activities. Traffic, mostly motor vehicles, contribute about 7000 tonnes of SPM per annum. Annual mean concentration calculated from the monthly mean data show a clearly increasing trend of SPM. In the SPARCENT (Space Science Division) site the annual means were 239 $\mu g/m^3$ in 1985, 265 $\mu g/m^3$ in 1986 and 275 $\mu g/m^3$ in 1987 (Ghauri *et. al.*, 1988).

A direct relationship between particulate constituents and street activity was identified in Beirut (Lebanon) by Chaaban *et. al.* (2000). The results indicate the seriousness of the air pollution problem that the transport sector can inflict on the society if appropriate mitigative measures are not implemented. Spot sampling in major street of the city showed particulate levels exceeding 300$\mu g/m^3$ an extremely high value when compared to WHO recommendations that set the range of 60-90μg/m3 as being acceptable in urbanized regions.

In **Mumbai** (India) SPM emissions have increased significantly in recent years and are projected to continue rising into the next century. In 1973, 67% of the total industrial SPM emission was caused by Power plants (NEERI, 1991b). SPM emissions attributable to transport have increased greatly. The transport sector doubled its share in the total estimated SPM emissions between 1970 (2%) to 1990 (4%). This proportion is likely to increase further with increasing motor vehicle traffic.

In **Kolkata**, has a severe SPM problem. Industrial sources account for 98% of the total (NEERI, 1991c). The high emissions and ambient concentration of SPM result from the excessive use of coal by industries. Emission indicates a 66% increase in SPM emissions between 1970-1980. The annual arithmetic mean concentrations in the 1990s were lower than in 1987 (268-453 $\mu g/m^3$) but still well above the WHO guideline.

Air pollution is one of the major problems faced by Delhi (India). Invariably, the transport, domestic, and industrial sectors are believed to be the main contributors towards the rise of ambient air pollution levels. The tremendous increase in the number of vehicles has contributed significantly to the increase in combustion of petroleum products. Petroleum consumption has increased by almost 400% in the last two decades (State of the Environment, TERI, p. 32). Among the industries, thermal power plants are the most prominent contributors to air pollution. Besides, there are 126 000 industrial units (small- and medium-scale units) in Delhi (Government of National Capital Territory of Delhi 1999) out of which 98000 units are categorized as non-conforming by the Master Plan of Delhi (Ministry of Environment and Forests 1997). The domestic sector, on its part, too contributes greatly to air pollution. According to the Government of National Capital Territory of Delhi 1999, the population of Delhi increased from 6.2 million in 1981 to 9.4 million in 1991.

The SPM (suspended particulate matter) in Delhi's air exceeds the permitted levels by over 100%. The RSPM (respirable particulate matter) levels were also found to be high (exceeding standards) at most of the monitoring sites in Delhi, reflecting that the prominent sources of RSPM are man-made (dust settled on the roadside, auto exhaust, coal combustion, *etc.*) RSPM constitutes about 40% of the SPM in the ambient air in Delhi (NEERI, 1991a).

To add to its woes, the pollution in Delhi is contributed, besides human-generated sources, by climate and natural sources. The region has a semi-arid climate that is often described as tropical, with extremely hot summers, heavy rainfalls in the monsoon months, and very cold winters. The summer months witness dust storms, while wintry evenings cause poor natural ventilation and high emission loads during peak traffic hours. It is only the monsoon season that reduces the pollution levels due to frequent washout of pollutants by rain.

In **Nagpur**, during August to December 1996, the average values of SPM ranged from 43-160 $\mu g/m^3$, 108-276 $\mu g/m^3$ and 59-186 $\mu g/m^3$ at Seminary Hill (control area), Itwari (crowded commercial/residential area with automobile and dust pollution) and MIDC (automobile and industrial gaseous discharge) area respectively. Highest concentration was recorded in November in Itwari area (Ninave *et. al.*, 2001). Khare *et. al.* (1996) reported SPM concentration in the industrial area of Agra was in the range of 173-973 $\mu g/m^3$.

The status of particulate matter having aerodynamic size less than 10 micron (PM 10) in four major coastal cities of India (Kolkata, Mumbai, Chennai and Kochi) at three sites in each city *viz.* industrial, commercial and residential is presented. It is found that Kolkata is most polluted city with reference to particulate matter among the four, whereas Mumbai is the second most polluted city (Chelani *et. al.*, 2002).

Sulphur Dioxide (SO_2)

The annual global emissions of sulphur dioxide stand at about 294 million tonnes of which of 160 million tonnes are anthropogenic (UNEP/GEMS, 1991). Total figures are calculated mainly from the estimates of production or consumption of fossil fuel, the sulphur content of fuel and mean emission factors. Fuel oil is the dominant source of SO_2 followed by coal and gas oil (Munday *et. al.*, 1989).

In 1990s the level of SO_2 in **London** was around 20-30 $\mu g/m^3$ (Laxen and Thompsom, 1987). In Los Angeles total anthropogenic emissions of oxides of sulphur were approximately 50000 tonnes, mobile sources accounting for 62% (SCAG 1991). The maximum hourly SO_2 measured at Hawthorne was 886 $\mu g/m^3$, which exceeds the WHO hourly concentration guidelines of 350 $\mu g/m^3$ (SCAG, 1991).

The largest source is **China**, which emits 18 million tonnes of SO_2 a year. The annual average sulfate levels in Southern Hemisphere ranged from 0.06-0.7g/m^3. Northern Hemisphere levels were typically 5 to 10 times greater than those measured in Southern Hemisphere. A pollution peak is seen in the Northern Hemisphere mid latitude regions where the highest level of industrial activity takes place. Average sulphate levels measured in Ireland during westerly wind conditions, which typically occur for 50-60% of the year, range from 0.9 to 1.2 g/m^3. These levels are principally determined by long-range transport of pollutants. During other periods sulphate level ranged between 10-12 g/m^3 (McGrovern and Wilson, 2000).

In **Germany** the annual SO_2 emissions are about 4 million tonnes, of which about 2.5 million tonnes are deposited within the country, and rest exported to neighboring countries, mainly Poland, Czechoslovakia and Sweden (UNEP/GEMS, 1991).

In **Dublin** city, SO_2 was measured and the average hourly concentration was 17.12 mg/m^3 and the main source was vehicular emission. Predicted concentration was 8.29 mg/m^3 (Reynolds and Broderick, 2000).

In **Bangkok** the major source of SO_2 is fuel oil used by vehicles, light and heavy industries (WHO/UNEP, 1992). Concentration of SO2 in ambient air is influenced by the meteorological conditions of the area, which favors the dispersion of industrial emissions outside the metropolitan Bangkok.

In **Beijing** the dominant source of SO_2 are vehicles, heavy industries and power plants. Industrial sources account for 87% of the total SO_2 emission. According to estimates, SO_2 emissions were about 526000 tonnes per annum in 1985. The city center had annual means over 100 μg/m^3 until 1989 and thus exceeded the WHO annual mean guideline (40-60 μg/m^3). The days with highest SO_2 (up to 250 μg/m^3) are between November and March and the annual average concentration of SO_2 was 39μg/m^3 (Krupnick and Sebastian, 1990).

In **Karachi**, total SO_2 emissions amount to about 77000 tonnes per annum, about 85% of which is emitted by fuel oil used by vehicles and power stations. Motor vehicles emit 2800, ships 1200 and aircraft 200 tonnes per annum (Ghauri *et. al.* 1988, Beg 1990a).

In **Mexico** City mean levels of SO_2 range from 80-200 μg/m^3, daily maximum being between 200-550 μg/m^3. No clear-cut trends have been noted between 1986-1991 (Gobierno de la republica Mexico, 1990).

In **Mumbai**, industrial sources account for nearly all SO_2 emissions. Following a 66% increase in SO_2 emission between 1970-1980, the rate of increase slowed significantly over the following 10 years. In 1990 total SO_2 emissions were around 157000 tonnes. This leveling off in the emission is largely due to the introduction of natural gas as a major fuel source, from the newly opened gas fields on the west coast (NEERI, 1991b)

In **Delhi** industrial sources and power stations are mainly responsible for most of the SO_2 emissions. The total anthropogenic SO_2 emissions are approximately 59000 tonnes per annum. SO_2

emissions from transport sources have increased and will continue to increase due to the increasing motor vehicle population (NEERI, 1991a)

In **Nagpur**, during August to December 1996, the average values of SO_2 ranged from 6-10, 6-19 and 6-19µg/m^3 at Seminary Hill (control area), Itwari (crowded commercial/residential area with automobile and dust pollution) and MIDC (automobile and industrial gaseous discharge) area respectively. Highest concentration *i.e.* 19 µg/m^3 was recorded in December at MIDC and Itwari (Ninave *et. al.*, 2001).

The changing levels of SO_2 along ten road transactions of Lucknow city (India) was done by Singh *et. al.* (1995). The road transaction around Alambagh (high traffic density) was highly polluted with SO_2 (202 µg/m^3), whereas the ambient air of road with lean traffic density (Vikramaditya Marg, Rana Pratap Marg and Nirala Nagar) were more or less clean as the levels of SO_2 (29, 33, 38 µg/m^3) were less than threshold level for SO_2 (60 µg/m^3).

Ambient air was monitored for a period of one year with respect to sulphur dioxide at three sites in an industrial city, Durgapur (West Bengal). It has been observed from the study that concentration levels was lower during July to October and higher during December to June, at all the three sites (Biswajit and Bhaumik, 2001)

Oxides Of Nitrogen (NO_X)

According to WHO's 1999 guidelines of Air Pollution Control (Air pollution) the concentration of NO_x was increasing in developed countries and developing countries. NO_x is increasing in countries such as Brazil, Chile, Hong Kong and India. The emissions are probably high in these countries because of the old and poorly maintained vehicles on roads. The ambient levels can be affected by several local characteristics such as tall buildings, meteorological conditions and photochemical conversions.

In **Western Europe** about 30-50% of man made emissions come from motor vehicles, and another 30-40% from power plants, mainly those fired by coal. It has been observed that 40% of the total NO_x emission in United States and 64% in Canada came from road transport. In Sweden about 30-40% of man made nitrogen compounds come from agriculture and forestry (UNEP/GEMS 1991).

Catalytic converters for road vehicles were first introduced in late 1970s and now are compulsory on new cars in Australia, Canada, Japan and United States and India. As a result emissions of NO_x from new cars in Japan were cut by 92% between 1972-78. During this period USA reduced emission by 75%. However these reductions in emissions have been largely offset by increase in the volume of traffic on roads. Several European cities including Amsterdam, Frankfurt and London have seen increase in NO_X, mainly from vehicles. Few European countries have stringent emission standards for NO_X emissions (UNEP/ GEMS, 1991).

Japan was the first country to set emission standard for NO_x from stationary sources. By 1986 Japan had installed more than 320 flue gas de-nitrification units in power plants and on other industrial units (UNEP/ GEMS, 1991). In Hiroshima NO_x is found having higher concentration in winter and lower in summer (Roy and Sukugawa, 2001).

In **Tokyo**, the man-made NO_x comes from high temperature combustion processes and the chemical industry (nitric acid and fertilizer production). Emissions were estimated to be 52700 tonnes per annum in 1985. Most of these emissions (35400 tonnes per annum or 67%) come from motor vehicle exhaust. Of the total NO_x emissions 18% comes from industrial sources 9.5% from domestic sources and 5% from ships and aircrafts (TMG 1989a). With respect to the NO_x emission from motor vehicle sources, diesel powered vehicles (especially cargo vehicles) are of particular importance although they contribute only about 20% of the total mileage travelled and contribute more than 50% of total NO_x emission. Conversely petrol powered vehicles that are equipped with exhaust pipe emission control devices contribute more than 50% of the total mileage travelled, but emit only about 21% of NO_x (TMG, 1989a).

For the whole of **Japan**, the 0.06 ppm standard was exceeded at about 25% of the automobile exhaust monitoring stations, which are located primarily in the large metropolitan areas such as Kanagawa, Osaka and Tokyo. In Tokyo during 1998, the 0.04-0.06 ppm daily average national standard was exceeded on several occasions (TMG, 1989b).

Transport particularly vehicles is the major anthropogenic source of NO_x in London. It accounted for about 75% of the total NO_x emission in 1984 in comparison to only 57% in 1975. The industrial/institutional emissions fell from 30500 tonnes (34%) in 1973 to 12200 tonnes (15%) in 1983 (Munday *et. al.*, 1989) London experienced the highest NO_2 concentrations in December 1991, with a maximum hourly average of 867 μg/m^3 (423 ppm). During the episode (12-15 December), NO_2 levels were in excess of 205 μg/m3 (100 ppm) for an average of 72 hours and over 600 μg/m^3 (300 ppm) for 8 hours. The episode was believed to be caused by motor vehicle emissions trapped during a period of cold calm weather (WSL, 1992).

In **Los Angeles** total NO_x emission in 1987 was over 4,40,000 tonnes per annum, 76% of which was attributable to mobile sources. According to USEPA (1991) Los Angeles is the only city in the US that has failed to attain the Federal NO2 NAAQS. The WHO one hour guideline of 400 μg/m3 corresponding to 0.22 ppm NO_2, was exceeded in 1990 at eight of the 24 stations reporting NO_2 data. The highest reported hourly value was 526 μg/m^3 (SCAG, 1991).

In **Moscow** the major source of oxides of nitrogen is electricity generation, which accounts for 70% of the total anthropogenic NO_x emission of 2,10,000 tonnes per annum. The contribution of motor vehicles to total anthropogenic NO_x emission does not exceed 19% in 1990. Daily mean NO_2 concentrations reached 100-150 μg/m^3, which is comparable to the air quality guideline recommended by WHO (150 μg/m^3) (CVGMO, 1986; MCGNPS, 1990,1991).

In **New York** man made NO_x emissions in 1985 was over 1,20,000 tonnes per annum New York city (NYC) and over 5,13,000 tonnes in the New York city metropolitan area (NYCMA). In 1990, NYC met the US NAAQS of 94 μg/m^3 annual mean with a value of 87 μg/m^3 (WHO/UNEP, 1992).

An oxide of nitrogen was monitored at a site within the central business district of Brisbane, Australia by Morawska *et. al.* (2002). Data was correlated with the traffic flow rate on the nearby free way with aim of investigating differences between weekday and weekend pollutant characteristics. The mean traffic flow rate on the freeway was 14.2×10^4 and 9.6×10^4 vehicles per weekday and weekend day respectively. The mean diurnal variations of particle concentration closely followed the traffic flow rate on weekdays and weekends. The effect of traffic on concentration of level of pollutant concentration in the vicinity of a major road carrying traffic of the order 10^5 vehicles per day is that about 50% increase in traffic flow rate will result in similar increase of NO_x.

In **Dublin**, NO_x was measured and average hourly concentration was 75.47mg/m^3 (Reynolds and Broderick 2000). Predicted concentration was 66.39 mg/m^3 by an emission inventory.

The mass concentration of NO_x was determined from four ground based monitoring sites in XiAn, China by Zhang *et. al.,* (2002). The samples were collected in four seasons during 1996 to 1997 and in July 1998. Annual average concentration of NO_x was 43 μg/m^3.

In **Cairo**, NO_x emissions from motor vehicles were estimated to be about 73,000 tonnes per annum in 1990, with two third of it coming from cars and one third from buses (WHO/UNEP 1992).

In **Seoul**, the emissions of NO_x increased markedly from 1984 to 1989. An increase in motor vehicle traffic has probably been the main factor for this change. The transport sector accounts for 78% of total NO_x emission. The total man made NO_x emissions in 1990 were estimated to be 2,70, 000 tonnes per annum in the Greater Seoul area and 1,30,000 tonnes per annum in Seoul (Rhee, 1991)

In **Nagpur**, during August to December 1996, the average values of NO_2 ranged from 4-12, 7-17 and 5-16μg/m^3 at Seminary Hill (control area), Itwari (crowded commercial/residential area with automobile and dust pollution) and MIDC (automobile and industrial gaseous discharge) area respectively. Highest concentration *i.e.* 17μg/m^3 was recorded in November and December at Itwari (Ninave et. al., 2001).

In **Mumbai**, transport was estimated to account for 52% of NO_x emissions in 1990 and increased by approximately 1,46,000 tonnes per annum by 2000. (NEERI,1991 b). Detailed vehicle emission inventories produced by the Department of Environment indicate that diesel vehicles (predominantly trucks) are the dominant source of motor vehicle derived NO_x in Mumbai.

In 1970 industry was the major source of NO_x (69%) in Kolkata, but now the dominant source is transport. Emission from industrial plant had increased to over 11000 tonnes per annum by 1980, which has stabilized since then. Transport emission have risen from an estimated 1,825 tonnes per annum in 1970 to 25,550 tonnes per annum in 1990 (NEERI 1991 c).

Pollutants dispersion behavior from the vehicular exhaust plume has a direct impact on environment. Results of the study by Chan *et. al.*. (2002) showed that mass concentrations of nitrogen oxides decrease along the central line of the vehicular exhaust plume in the downstream distance. The dispersion process can be enhanced when the vehicular exhaust tailpipe velocity is much larger than the wind speed. The oxidation reaction of the NO plays an important role when the wind speed is large and the vehicular exhaust exit velocity is small, which leads to chemical reduction of NO_x, and the formation and accumulation of NO_2 in the exhaust plume.

Carbon Monooxide (CO)

Carbon monoxide is colorless, odorless and taste less gas that is poisonous and lethal. CO is a by-product of incomplete combustion. It is produced when inflammable fuels such as natural gas, propane gas, heating oil, kerosene, coal, charcoal, gasoline or wood burn with insufficient oxygen.

In the **Unites States**, the number of vehicles increased four fold between 1940 and 1976, while annual CO emissions rose from 73 to over 100 million tonnes (UNEP/GEMS, 1001). Between 1977 and 1985, the number of vehicles grew by 25 %, but CO emissions fell from 80 million to 60 million tonnes annually due to control measures.

In the United States, emissions from vehicles fell by 33 % between 1980 and 1989. In the Netherlands and New Germany, emissions fell by 32 and 37 % respectively, between 1975 and 1985.In Britain, Ireland and several East European countries, however, emissions grew over the same period by 5.2 % because of increased traffic congestion (UNEP/GEMS, 1991).

Carbon monoxide concentrations touch peaks in winter during high pollution episodes. These episodes tend to correspond with ground based temperature inversions and low wind speed which trap vehicle emissions close to the ground.

Dealing with CO emissions in developing countries will be more complex. CO emission in Bangkok was 120000-160000 tonnes per annum in 1980 and could grow up to 420000 tonnes per annum in 2000(Faiz *et al.*, 1990).

In **Mumbai**, estimated emissions increased form 69000 tonnes per annum in 1970 to 1888000 tonnes in 1990-1991and 255500 tonnes in 2000.Most of the increase is attributed to motor vehicle transport which was estimated to be responsible for 89 % of the total CO emissions in 1990 (NEERI, 1991b).

In **Delhi**, CO emissions increased form 140000 tonnes per annum in 1980 to 265000 tonnes per annum in 1990 (NEERI ,1991a).

Ozone (O_3)

Ozone is a natural constituent of the atmosphere, and life on earth depends on its presence. Surveys have revealed that precursor emission generated in London causes increase downwind O_3 concentration between 38 and 154 $\mu g/m^3$ (18-72 ppb) within a few hours (Upto 10 hours).

Ground level ozone occurs naturally but level can be increased as a result of the presence of other pollutants. Tropospheric ozone or ground level ozone originates from a series of reactions between hydroxyl radical, nitrogen oxides, methane (CH_4), Carbon mono oxide (CO), Oxygen (O_2) and Volatile organic compound (VOCs).

Ozone episodes in which concentration rise substantially above background levels occur in summer heat waves when there are long hours of bright sunlight temperature above 20 ^{0}C and little or no winds.

For many years it was believed that O_3 in the troposphere was more or less inert. However relatively recent advances in measurement techniques have demonstrated that there are pronounced spatial and temporal variation in tropospheric O_3 concentrations. This variability is partly due to natural processes, but anthropogenic sources of other pollutants also effect the distribution and abundance of O_3. Recent studies have shown that O_3 induced physiological and biochemical disturbances may persist for many months after exposure to the pollutant has terminated and possibly even persist from one year's summer exposure to the next.

Once formed ozone can persist for several days and can be transmitted long distance at ground level. Ozone can affect human health and can damage plant and crop. WHO also recommend guidelines for ground level ozone in terms of 1 hr mean concentration between 76 and 100 ppbv and on 8-hrs mean concentration between 50 & 60 ppb.

The developing world may be a minor emitter of ozone precursors but the climatic conditions are favorable for its production in many countries. Concentration of ozone is higher in the warmer months due to higher temperature and greater stagnation. Root cause of high O_3 emission in developing countries is the use of fossil fuel use in industries and automobiles.

Los Angeles has the most serious O_3 problem in the USA (USEPA, 1991). In New York, annual emissions of reactive organic compounds (ROC's) in 1985 totaled 216000 tonnes in New York City, and over 870000 tonnes in NYCMA. Areas of O_3 maxima in NYCMA are "downwind" of the city proper depending on wind direction, *e.g.*, in New Jersey or Connecticut.

In **Tokyo**, the annual mean concentrations of photochemical oxidants have decreased since the late 1960s (TMG, 1989b). While the annual mean concentrations of oxidants were as high as 0.035-0.045 ppm (70-90 $\mu g/m^3$ expressed as O_3) in 1968-1970, the levels dropped to about 0.02 ppm (40 $\mu g/m^3$) in 1978.

In **Bangkok**, O_3 is not a problem because of the year–round monsoon that prevents build-up of pollutants. The highest concentrations were found in the March- May season, when solar radiation is strongest. The maximum 1-hour mean O_3 measured was around 100 $\mu g/m^3$, which is below the WHO guidelines value of 200 $\mu g/m^3$ (WHO/UNEP, 1992).

In **Beijing**, O_3 was 158-262 $\mu g/m^3$ (hourly) in June 1986 and 133-320 $\mu g/m^3$ in July 1987, which was maximum (Tang *et. al.*, 1988). These high concentrations of O_3 are due to the high NO_x emission. O_3 level in Mexico city are exceptionally high (DDF, 1989; Gobierno de la Rebublica Mexico,1990). The annual mean O_3 fluctuated around 200 $\mu g/m^3$ with minimum value 100-150 $\mu g/m^3$ and maximum value of 300 and 400 $\mu g/m^3$.

Air Quality Status in Major Indian Cities (Source: Annual Report 2002-2003, CPCB)

Central Pollution Control Board initiated National Ambient Air Quality Monitoring (NAAQM) programme in the year 1984 with 7 stations at Agra and Anpara. Subsequently in 1998-99 the programme was renamed as National Air Monitoring Programme (NAMP). The number of monitoring stations under NAMP has increased, steadily, to 295 by 2000-01 covering 98 cities/towns in 29 States and 3 Union Territories of the country. Under NAMP, four air pollutants *viz.*, Sulphur Dioxide (SO_2), Oxides of Nitrogen as NO_2 and Suspended Particulate Matter (SPM) and Respirable Suspended Particulate Matter (RSPM/ PM_{10}), have been identified for regular monitoring at all the locations. The monitoring of pollutants is carried out for 24 hours (4-hourly sampling for gaseous pollutants and 8-hourly sampling for particulate matter) with frequency of twice a week, to have 104 observations in a year.

It has been observed that air pollution problem is serious mainly due to high vehicular population in seven major cities in India namely Ahmedabad, Bangalore, Chennai, Delhi, Hyderabad, Kolkata and Mumbai. An attempt has been made to address the problem of air pollution in these cities. Estimates are made of air pollutants load coming from vehicles. Status of air pollutants is also established to find the air pollutants that are exceeding the air quality standards.

Urbanization in India is more rapid around national capital and state head quarters. Over the years, these cities have become a major center for commerce, industry and education. Increase in population, both endemic and floating, increase in industrial activities, vehicular population *etc.* have led to a number of environmental problems, one of them being air pollution. Enormous increase in number of vehicles has resulted in increased emission of air pollutants from motor vehicles.

SO_2, and NO_2 levels are within NAAQS (Annual average) in Ahmedabad, Bangalore, Chennai, Delhi, Hyderabad and Mumbai.

Short-term violation of NO_2 is observed especially during winter months in Mumbai and Delhi.

Most critical form of pollution is Respirable Suspended Particulate Matter prevailing at all the seven cities.

SPM levels are also exceeding the NAAQS (Annual average) in Ahmedabad, Bangalore, Delhi, Hyderabad, Kolkata and Mumbai in residential areas. The reason for high particulate pollution may be vehicular emissions, resuspension of dust, commercial and domestic use of fuel *etc.* In Ahmedabad emission from power plants and industries located in industrial areas may also contribute to particulate matter pollution. In Mumbai, emission from power plants and oil refinery may also contribute to particulate matter pollution.

Chapter - 5

Effect of Air Pollutants on Biotic and Abiotic Components of Ecosystem

Effect of Air Pollutants on Plants

Elements of those air pollutants which accumulate in plants (S from SO_2, heavy metals from anthropogenic emissions *etc.*) may be used for specific bio-indication of possible effects caused by these pollutants, and for mapping their spatial and temporal distribution. Measuring accumulated components in plants rather than directly in the environment, provides the dual advantages of an integrated rather than a momentary value, and of a biological content rather than a mere concentration (Stocker and Gluch, 1990).

Plant response has been divided into avoidance and tolerance mechanisms. Avoidance can be considered as a system, which prevents the stress from acting on the plant, whereas tolerance is the physiological change in metabolism, which reduces the intensity of the protoplast, or allows stress damage to be repaired (Larcher, 1994)

Effect of air pollution on plants has long been known (Bobrov, 1955; Solberg and Adams 1950). Various changes induced by air pollutants in plants with respect to morphological and anatomical (Sharma 1977, Yunus *et.al.* 1979, Garg and Varshney 1980, Mishra 1982) physiological (Dugger *et.al.* 1962, Lee 1965, Crittenden and Read 1979) and biochemical (Cracker 1972, Tringey and Reinert 1975, Malhotra and Hocking 1976) characteristics have been recorded.

Physiological Changes

Air pollutants such as SO_2, NO_x and HF influence plant physiology on their own or in association with certain climatic factors (*e.g.* drought). SO_2 affects needles/ leaves of trees and therefore hinders the assimilation process (Linzon1978, Mass 1987, Rennerberg and Herschblach, 1996). Conifers are especially sensitive to such conditions and continue to reduce CO_2 uptake in young spruces (Keller 1980). The reduced photosynthetic activity in turn retards cambial activity and consequently wood production. Concentration of SO_2 is more decisive than duration of exposure in affecting photosynthesis (Guderain, 1998). Trees are considered as active bio-indicators of air pollution as they react sensitively to all interfaces (Braun and Lewark, 1986).

Senescence is one of the physiological responses of plants, which is accelerated due to air pollution (Kohert *et.al.* 1986). Overt manifestations of plant senescence are the gradual disappearance of chlorophyll and concomitant yellowing of leaves, which may be associated with the consequent decrease in the capacity for photosynthesis (Woolhouse, 1986).

Morphological and Anatomical Changes

Different levels of air pollution input produced different sets of harmful effects. Although the air pollution level at Varanasi reduced the plant height, basal diameter, canopy area, leaf area and total plant biomass of *Cassia carandas,* this species retained a major fraction of its photosyntheate to above-ground plant parts where foliage assumes predominance (Pandey 1997).

The sensitive plants indicate injury symptoms due to air pollution. The phenotypic symptoms of sensitive plants due to SO_2 and NO_x are specific to the type of pollutant, concentration and exposure time such as 0.70ppm (18.20 $\mu g/m^3$) of SO_2 for 1 hr produces Interveinal Necrosis, 0.18ppm (468 $\mu g/m^3$) for 8 hr produces blotches, 0.008-0.017ppm (21.44 $\mu g/m^3$) during growing season produces Red brown dieback or Banding in pines; NO_x 20ppm ($38x10^3$ $\mu g/m^3$) for 1 hr, 1.6-2.6ppm (3000-5000 $\mu g/m^3$) for 48 hr.; 1ppm (1900 $\mu g/m^3$) for 100 hr produces interveinal necrotic blotches similar to those of SO_2 (Smith, 1981a and b).

Biochemical Changes

Two important mechanisms of plants, photosynthesis and protein production, are decreased due to air pollution (Amundson *et. al.* 1986,

Khan *et. al.*, 1990). *Cassia tora* shows low chlorophyll content in normal environment whereas *Xanthium strumarium* and *Argemone mexicana* show high value in polluted environment (Jain, 1992). The chlorophyll content of wheat crop is reduced to 8.9% due to coal dust has been reported by Pandey *et. al.*, 1992.

Chlorophyll content of five plant species was analyzed by Madhumanjari *et. al.* (2000). A decrease of 33, 23 and 12% was observed for *Nerium, Tabernaemontana* and *Boerhaavia* respectively whereas *Amaranthus* and *Cephalandra* recorded increase of 26 and 16% than control.

Variations In Peroxidase Activity

Air Pollution Marker Enzyme

Peroxidase enzyme belongs to the class of oxido-reductase, occurs in a wide variety of trees and shrubs and has been extensively studied in attempts to develop a method for early identification of chronic injury from air pollution. Air pollution can produce an H_2O_2 stress on plants (Puccinelli *et. al.*, 1998). Peroxidase is also involved in the detoxification process of H_2O_2 produced in injured plant cells. Studies of Keller (1974) first reported the possibility of using leaf peroxidase activity as a marker to evaluate air pollution. In urban environments different studies reported that leaf peroxidase activity was enhanced in places which were more exposed to air pollution caused by intense car traffic (Mannes *et. al.*, 1986; Mannes *et. al.*, 1989; Mannes, 1987, Bragaloni *et.al.*, 1993; Bragaloni, 1995). Air pollution induced guaiacol activity in leaves of trees was verified as marker to evaluate air quality. According to a study conducted by Puccinelli (1998), in Turin, air pollution increased peroxidase activity in leaves was verified on 530 plants, both evergreen and deciduous broadleaf. Peroxidase activity of the trees in the urban environment was higher than those from non-urban. Peroxidase activity was highest in December (winter) and low in August. Karjalainen *et. al.*, 1992 reported peroxidase, a specific indicator of SO_2 and NO_2 pollution. Acute levels of SO_2 particularly combined with NO_2, strongly induced peroxidase activity. Peroxidase activity varies with the species of tree, season and concentration of the gas. The changes in the peroxidase activity are accompanied by the changes in catalase activity (Nandi *et.al.*, 1984).

The mechanism of the injurious action of SO_2 on enzymes can be explained in part by the similarity of sulphate with some other anions and in part by its competition with CO_2 in metabolic processes. Conversely however, peroxidase from the needles of spruce, despite being an iron-containing enzyme is not inhibited rather its activity increases when are subjected to stress (Khan *et. al.*, 1990).

As SO_2 enters the leaf it is converted to bisulphite and sulphite. Bisulphite is photo oxidized to less toxic sulphate and super oxide radical. The superoxide radical can be dismutated by super oxide dismutase enzyme to H_2O_2 and O_2^-, H_2O_2 is converted to H_2O and O_2 by peroxidase and catalase. Toxicologically O_2^- and H_2O_2 are considered particularly important as precursors for the highly reactive and destructive hydroxyl radical. Cell injury can be severe if even a small portion of the superoxide is converted to OH^-. Prolonged exposure to metal concentrations may enhance peroxidase activity (Grunhagel *et. al.*, 1981).

In Scots pine, young needles have an active Peroxidase in the epidermis and walls of sclrenchyma cells. In mature needles the activity occurred in the protoplast. Localization of the enzyme undergoes no change depending on whether the tree is growing in clean environment or in environment polluted with SO_2 and HF (Przymusinski, 1980). In sensitive trees the enzyme activity is higher in mesophyll tissue, while in resistant tree the activity is higher in the epidermis, hypodermis and resin canals. The activity was drastically high in the cells of hypodermis and endodermis of tree resistant to SO_2 when they were exposed to the gas. These tissues in the needles may have significance in the process of detoxification.

Plants

Bioindicators of Pollution

The leaves of *Mangifera indica* and *Delonix regia* were found to be most efficient dust interceptor among the various plant species investigated in Singrauli industrial belt in the Sonebhadra district of Uttar Pradesh where several thermal power plants and open-cast coal mines are operating. The leaf area of different plant species was considerably reduced at sites receiving higher gaseous and particulate load. The photosynthetic potential of fruit trees and roadside plantations, measured in terms of total chlorophyll were considerably reduced at the heavily polluted sites and as a result, the leaf biomass accumulated per unit area also decreased. The foliar starch content of the tree species was significantly reduced. Plants growing near the power plants accumulated higher concentrations of total S, SO_4, Ca, K & trace elements. Evergreen trees were found to be more sensitive during summer, whereas deciduous trees and shrub were more sensitive during winter. Plant response study indicates that *Mangifera indica, Delonix regia* and *Eucalyptus hybrid* are potential bioindicators around S emitting sources. *Mangifera indica, Cassia siamea* and *Eucalyptus hybrid* are bio accumulator of sulphur. The phytotoxic role of SO_2 was confirmed through simulation experiments. Lower concentrations of SO_2 were found to be beneficial for plant growth, while higher concentrations in combination with fly ash caused adverse effects on the growth, metabolism and productivity of selected crop plants (Rao, 1989).

Deciduous plant species are more susceptible to pollution stress than the evergreen species (Tiwari and Bansal 1993). *Mangifera indica* (Agarwal and Najma 1989) and *Ficus benghalensis* (Yunus *et. al.* 1991) are tolerant to dust.

Particulate Matter

Many pollution-related adverse effects are closely related to the presence of particulate matter, fine (<2.5 μm), and even more, to ultrafine airborne particles (<0.1m μm). These particles usually contain most of the trace elements and toxins and, due to their higher diffusion coefficients, penetrate deeper. This response is hypothesized to depend less upon the mass of the particles than on their number and size distribution.

In highly polluted urban environments that are significantly influenced by motor vehicle emissions, number, concentrations of particles may exceed 10^5 $cm^{-3.}$ Harrison *et. al.*. (1999) found that particle number concentration was 7.5 times higher than the background near a busy road in Bristol, UK. Most of the particles emitted by engines are in the ultrafine range (Kittelson, 1998) and consequently, it is not surprising that in urban environments not directly influenced by industrial emissions, over 80% of airborne particles are in this size range (Morawska *et. al..*, 1998b; Shi *et. al.*, 2001).

The particles emitted by motor vehicles are mostly black carbon soot. They range in size from 0.02 to 0.13 μm from diesel engines (Morawska *et.al.*, 1998a) and 0.04 to 0.06 μm from petrol engines (Ristovski *et.al.*, 1998). While most light vehicles operate on petrol, most heavy vehicles such as buses and other transport and construction sector vehicles generally use diesel engines emit less organic compounds than petrol engines, they are higher emitters of particulate matter. Through fundamental improvements in engine design and the introduction of catalytic converters and unleaded fuels, technological advances in recent years have led to a marked reduction in gaseous and particle mass vehicular emissions. However, these advances have not resulted in a decrease in the emission rates of ultrafine particles. Quite to the contrary, the changes in the combustion process have in fact resulted in increased emission of these particles (Bagley *et. al.*, 1996; Kittelson, 1998; Hall *et. al.*, 1998.

Effect On Plants

Particles can produce a wide variety of effects on the physiology of trees. Heavy metals and other toxic particles have been shown to accumulate, causing damage and death of some species (Fatoki and Ayodele, 1991; Sturaro *et. al..*, 1993; Singh *et. al..*, 1995; Alfani *et. al..*, 1996). With further regard to the phytotoxicity of heavy metals, although many concentrations encountered are sufficient to cause ill-health in humans, they are often enough to cause only minimal toxic effects in trees (Lin and Schuepp, 1995).

Spatial variation in dust deposition and yield were correlated. *Cicer arietium, Glycine max* and *Cajanus cajan* in cement factory showed reduction in plant height, total chlorophyll and fresh dry weight of plants growing at polluted sites as compared to control sites (Varshney *et. al.*, 1997).

Heavy loads of atmospheric particles such as those which can occur close to unpaved roads and open cast quarries, also result in the occlusion of stomata, and decrease in the efficiency of gaseous exchange. The resultant crust of particles on leaf and bark surfaces disrupt other physiological processes such as bud break, pollination, light absorption and reflectance (Farmer, 1993).

Higher SO_2 and NO_2 levels during the winter cause greater pollution damage at this time of the year to plants (Whitmore and Mansfield, 1983).

Sulphur Dioxide

Sulphur containing air pollutants are deposited in the terrestrial ecosystem *via* 3 mechanisms, dry deposition, wet deposition, and occult deposition. Dry deposition is the process by which pollutants are transferred from the air by impaction and uptake. Wet deposition is the process by which pollutants are removed by precipitation and are transferred to the ground *via* the gravitational process (rain, snow). The third route occult deposition includes pollutant transfer in mist, fog, and cloud water.

Sulphur is available to plants as sulphate in the soil to the roots and or as gaseous sulphur compounds in the atmosphere to the shoots. Despite significant reduction of emission from the anthropogenic sources (Whelphale, 1992), SO_2 is still the most important atmospheric sulphur compound taken by vegetation. Sulphate influx in roots is mainly an active carrier mediated process (Clarkson *et. al.*, 1993) controlled by the sulphur nutrition of the plant (Herschbach and Rennerberg, 1994), influx of sulphur is mainly a diffusive flux through the stomata dependent on stomatal aperture (Rennenberge & Pole, 1994). Thus, avoidance of SO_2 influx into plants can only be achieved at the cost of reduced photosynthesis and water – vapour exchange hence reduced growth.

Sulphur dioxide that has entered the gas phase of a leaf rapidly dissolves in the aqueous phase of the cell wall and reacts with water to form bisulphate. The equilibrium of this reaction is far on the side of bisulphate at the p^H of the apoplastic space (Rennberg and Polle, 1994). Plasmalemma is impermeable to bisulphate or sulphite, but SO_2 (aq) may pass plasmalemma by diffusion and therefore by a relatively slow process. Rapid metabolic conversion of sulphite to sulphate may be achieved by apoplastic sulphite oxidase activity (Pfanz *et. al.* 1990). Because this conversion requires molecular oxygen, hydrogen peroxide and monophenols, the reaction is catalyzed by apoplastic peroxidases.

Sulphate produced from sulphite in the apoplastic space will enter a large apoplastic sulphate pool (Wolfenden *et. al.*, 1991) transported to the leaf with the transpiration stream. From this pool sulphate derived from the SO_2 will enter the symplasm by active carrier mediated transport through the plasmalemma (Clarkson *et. al.*, 1993). Thus SO_2 exposure may result in enhanced apoplastic sulphate availability and elevated sulphate influx into leaf cells.

Exposure of foliage to atmospheric SO_2 results in the transport of SO_2 derived sulphur to the roots in the form of sulphate and organic sulphur compounds. Atmospheric SO_2 interacts in plant with processes involved in the regulation of sulphur nutrition at the whole plant level, but no diagnostic tools have been developed to benefit from this observation. SO_2 and its metabolic products affect numerous physiological processes in plant cells, including stomatal aperture and its control, carbon balance and its allocation of photosyntheate, chloroplast functions and antioxidant defence mechanisms (Bonneau *et. al.,* 1995).

There is however hope for the survival of some plants because plants have varying tolerance to SO_2 (Tsukahara *et. al.,* 1985) and some plants can adjust to air pollution (Wilson and Bell, 1985). Certain plants can even show stimulated growth, as for example *Sphagnum* and epiphytic lichens in response to SO_2 and NO_2 (Lee JA *et. al.,* 1990) become more abundant in an increasing air pollution gradient (Oksanen *et. al.,* 1990). Light can reduce damage caused by SO_2 (Olszyk and Tingey, 1984).

Effect On Plants

Sulphur dioxide lowers net photosynthesis; leaf respiration and transpiration of *Pinus banisiana* and *P cordata* (Amudson *et. al.,* 1986), and assimilation rates are lowered by SO_2 fumigation in Pinus species and other trees (Addison *et. al.,* 1984). Thus both primary metabolism (Koziol and Whaltey, 1984; Darral 1986) and secondary metabolism are altered (Juttner 1988, Katoh *et. al.* 1989). Direct damage done to plant tissue can destroy the plants ability to defend its cell against pollution stress (Lechowicz, 1987).

SO_x cause adverse impact on vegetation including forest and agricultural crops. Studies in United States and elsewhere have shown that plants exposed to high ambient concentrations of SO_2 may loose their foliage, become less productive or die prematurely. Some species are much more sensitive to exposure than others. Plants in the immediate vicinity of emissions sources are more vulnerable. Studies have shown that the most sensitive species of plants begin to demonstrate signs of visible injury at concentrations of about 1,850 $\mu g/m^3$ for 1 hour, 500 $\mu g/m^3$ for 8 hours, and 40 $\mu g/m^3$ for the growing season (NAPAP 1990).

The effect of SO_2 pollution in field conditions was studied examining the transplanted saplings at the polluted site of Indraprasth Thermal Power Plant, New Delhi and comparing the results with the plants growing at relatively non-polluted site of JNU. Exposure to SO_2 pollution adversely affected the plant as indicated by foliar injury in the form of chlorosis and necrosis and reduced phytomass (Shahare and Varshney, 1994).

Plants growing in the SO_2 polluted area showed a marked decrease in total chlorophyll content, like Ficus showed a 25% decline in the total chlorophyll content (Agarwal and Tiwari, 1996). Four-month-old Polyalthia plants exposed to 0.12 and 0.25 ppm SO_2 for 4 hrs daily for three weeks, showed no characteristic SO_2 foliar injury symptoms. There was accumulation of reducing sugars with loss in starch content. Chlorophyll and free amino acid contents showed constancy in their amounts while protein contents were significantly reduced (Kumar and Jayashree, 1999).

Oxides Of Nitrogen

The increasing trend of NO_x concentration in urban areas has gained much attention recently as problem of atmospheric NO_x pollution in cities is increasing. Plants have a large capacity for NO_x uptake and tolerance, and thus purify atmosphere. Comparisons of NO_x uptake rates between different types of plants was done by Okano *et. al.*, 1988. Morikawa *et. al.* (1993) used N_{15}-labelled NO_2 to fumigate 14 species of both wild and cultivated herbaceous and woody plants, and found difference of more than 560 fold in the efficiency of different species to assimilate NO_2.

Effect on Plants

NO and NO_2 produce numerous physiological changes similar to those found with other oxidants. The response includes membrane damage due to inhibition of lipid biosynthesis and lipid per-oxidation and reduction in net photosynthesis (Welburn 1990).

Evergreen plant *Mimusops elengi* showed decrease in photosynthetic pigment under NO_2 pollution conditions (Tiwari and Bansal 1993). Reduction in photosynthetic pigment of Ficus relegiosa was observed by Imiwari *et. al.* (1992) under NO_2 pollution conditions. One of the major effects of increasing atmospheric deposition of N is believed to be an increase in the sensitivity of plants to winter stress.

Ozone

The ozone enters into plant leaves through stomata. During exposure to ozone, stomata on the resistant cultivars exhibit partial closure, whereas those on the sensitive cultivars did not show closure.

However, the ozone may react directly with protein or lipid component of membrane, or free radical produced due to ozone activity within the substomatal cavity or may directly react with membrane component. Ozone produces variety of effects such as leakage of solutes and loss of turgour control, change in chloroplast ultra structure and disruption of electron transport. Ozone caused a substantial decrease in the light saturated rate of photosynthesis, which occurred together with a decrease in stomatal conductance.

Ozone and other gaseous pollutants entering through the stomatal cavity dissolve in extra cellular fluid and reach with biological compounds to produce cationic species of free radicals. Free radicals interact with proteins and lipids in the cell wall and membrane leading to oxidation and lipid peroxidation and initiation of chain reaction giving rise to more free radicals followed by increase in cell permeability. This all can lead to severe metabolic disorder and damage to genetic material.

These gaseous pollutants *viz.* ozone, sulphur dioxides *etc*, destroy the ultra structural organization of leaf cells. Thylakoids swell, curl and then disintegrate. The cytoplasmic and plastid matrix decrease in number.

Pollution in the USA has been attributed to photochemical oxidants of which ozone is the most important component. The dissolution of ozone in the water leads to decomposition products such as super oxide, peroxyl and hydroxyl radical at physiological pH. However only hydroxyl radicals are detected especially in presence of phenolics such as caffeic or ferulic acid. The reaction of ozone with substrate that yields H_2O_2 can increase the O_2 radicals and consequently leads to generation of highly reactive hydroxyl radical's OH- radicals may attack membrane with subsequent fixation of the damage by the formation of peroxy radical. This involves members derived organic oxygen radicals. O_2^- serves only as a chain propagator. The damage caused by ozone include oxyradicals especially prevalent in chloroplasts where their concentration may be enhanced during illumination, resulting in light induced injury to susceptible chloroplasts. Starch and sugar in carbohydrate loaded chloroplasts, on the other hand might mitigate the injury, Ozone and secondary products can react with a wide range of bio-organic molecular and inorganic cell constituents. High ozone concentration cause general destruction to the cells. Chronic low –level concentrations often present in ambient air can have more specific effects.

The most widely observed chemical change induced by ozone is the destruction of chlorophyll (Guderian *et. al.* 1985), chlorophyll-b is more sensitive than chlorophyll –a and carotenoids are less sensitive to ozone then chlorophyll. The light saturated rate of CO_2 uptake decreased significantly when plants were exposed to elevated levels of ozone.

The long –term exposure of plant to ozone results in the loss of chlorophyll content, which appeared very late and did not provide a satisfactory interpretation for the initial reduction in the rate of photosynthesis.

In the chloroplasts the election transfer from ferrodoxin to oxygen and the subsequent formation of H_2O_2 *via* super oxide dismutase necessitates a pathway to prevent build up of H_2O_2.

VEGETATION IMPACT STUDIES IN DIFFERENT COUNTRIES

In **China** the air pollutants are of greatest concern relating to vegetation impacts of SO_2, NO_x and SPM. Most SO_2 emissions are associated with the eastern part of the China where population density and industrial activity is high. Pollution by NO_x is not considered as a serious problem as SO_2. Observed impacts include the development of red and brown necrotic lesions and a thick layer of black dust on the surface of the leaves after several days without precipitation. Other symptoms include delayed sprouting and accelerated senescence. Other tree species such as peach, cherry and citrus also exhibited the signs of air pollution induced injury with inhibited or shortened flowering period, early leaf abscission and premature dropping of fruit (Zheng and Chen, 1991).

Air pollution impacts have also been reported in industrial complexes in China including power stations, steel plants and cement factories. In the human province air pollutants from coal burning power stations resulted in severe impact on local agriculture causing in some 100% losses in yield. Citrus trees growing in the vicinity of power station showed reduction of about 55% in leaf CO_2 assimilation rates and 50% reduction in fruit number per tree compared with those growing in clean area (Boa and Zhu, 1997. In China most agricultural land (including that used for fruit and vegetable production) is located around cities and industrial areas at low elevation. As such fruit trees and vegetables are considered at the greatest risk from air pollutants.

Air pollution in **Taiwan** has become an important environmental issue. A field survey performed by Lin and Yang (1996) found that many crops exhibited O_3 injury symptoms in South Taiwan.

In **Pakistan**, Lahore and Karachi are two urban areas that have experienced severe reduction in air quality over the last decades. Monitoring of NO_2 concentrations around Lahore in 1993-1994 recorded elevated NO_2 levels, decreasing from city centers (weekly means of 70 $\mu g/m^3$) to rural areas (weekly means of 10 $\mu g/m^3$). In Lahore controlled experimental investigations using OTCs have investigated the impact of ambient O_3 and NO_x on the growth of local crop plants such as wheat, rice, chickpea, mungbean and soyabean crops in comparison with those growing in pollution free air. The damage caused by the exposure of plants to ambient air pollution includes reduced number of tillers, shoots and leaves, accelerated leaf senescence and yield reduction of between 23 and 47% (Wahid, *et. al*., 1993.).

The high level of air pollution has been reported in urban industr centers of Cairo and Alexandria in Egypt. It has great impact becaus the area is the primary agrarian region situated near the famous Nile river. Air pollution ultimately polluted the river, which is the main source of irrigation for cultivation. Ali *et. al.* (1993) reported instances of visible injury on clover/berseem (*Trifolium*) and Egyptian Mallow (*Malva parviflora*) plants growing close to the industrial complexes at Shoubra El Khaima. Mean pollutant concentrations for SO_2, NO_X and TSPM between November 1987 and January 1988 was 160, 88 and 680 $\mu g/m^3$ respectively. Visible injury included necrosis and chlorosis with 60% and 54% of clover and Egyptian Mallow leaves injured respectively.

In **South Africa** there are a number of locations where air pollution is perceived to be a problem. SPM emissions from household coal and wood burning are a concern in urban areas. Due to high SO_2 emissions, the commercial forests have been most extensively studied for air pollution impacts to vegetation in South Most trees exhibited some type of chlorosis, which appeared to be correlated with the potential pollution impact at the site (Olbrich, 1990).

The most serious air pollution in Mexico occurs in the vicinity of Mexico city associated with high O_3 and SPM concentrations. Other contributors to the air pollution problem include the Mexican oil producing areas located in southern Mexico where high SO_2 concentrations along with other pollutants are thought to be causing serious damage to local tropical vegetation and crop species. SPM is an important air pollutant in the valley of Mexico, as well as some other parts of the country.

In **Brazil**, the pollution varies significantly across the country. Pollutant concentrations are highest in the southern and southeastern regions of Brazil . These regions are associated with the location of large urban and industrial activities. The most renowned industrial complex of air pollution is located at Cubatao in the state of Sau Paulo. Active and passive bio-monitoring has been performed within the region to identify the damaging pollutants. SO_2 (annual mean concentrations of SO_2 was 26 and 18 $\mu g/m^3$) was found to be the most damaging pollutant. An increase in foliar concentrations of Sulphur was found in plants indicating that SO_2 was inducing vegetation damage (Klumpp *et. al.* 1996).

In **India** concern has been raised over air pollutants concentrations of SO_2, NO_x and SPM. The annual average SO_2 ranged from 10-120 µg/m^3 across the country (Agarwal *et. al.*, 1991). Annual average NO_2 concentrations ranged from 10-90 µg/m3 with especially high concentration around metropolitan cities. In general northern and western part of the country experience high levels of air pollutants as compared to south and eastern parts. Adverse impact of air pollution on plants around industrial sources and metro cities has been reported for many parts of India. Field studies have identified SO_2 as the important air pollutant contributing to yield reductions up to 50% in agricultural species growing in the vicinity of thermal power plants where SO_2 concentrations of 75-135 µg/m^3 were recorded (Agarwal Press comm.). Studies on three varieties of *Oryza sativa* growing under ambient conditions near fertilizer plant in Baroda, Gujrat have shown that high concentrations of SO_2 and NO_2 (mean maximum values of 144 and 210 µg/m^3 respectively) induced significant injury. Damage included reduction in panicle length, dry weight of filled grain and grain yield compared to a relatively pollution free site (Anbazhagan *et. al.*, 1989).

The vegetation of the Nagpur city, India is exposed to dust pollution and chronic concentrations of gaseous pollutants, which may effect the biochemical makeup, and tolerance capability of plants to air pollution (Ninave *et. al.*, 2001). The total chlorophyll content (mg/g dry weight) in the leaves of *Azadirachta, Pongamia, Bougainvillaea* and *Polyalthia* from Seminary Hill (Control area, where SPM, SO_2 and NO_x ranges from 43-160, 6-10 and 4-12 µg/m^3 respectively) was 2.97, 1.74, 3.73, and 4.31 respectively. In the Itwari area (residential cum commercial area) with SPM, SO_2 and NO_x values in between 108-276, 6-19 and 7-17 µg/m^3 respectively showed chlorophyll level 2.50, 2.88, 3.44 and 1.74 in the leaves of *Azadirachta, Pongamia, Bougainvillaea* and *Polyalthia* respectively. In commercial cum industrial area MIDC (SPM- 59-186, SO_2 - 6-19, NO_x- 5-16 µg/m^3) the chlorophyll content *Azadirachta, Pongamia, Bougainvillaea* and *Polyalthia* was 3.94, 2.85, 4.57 and 4.44 respectively. Chlorophyll content in the leaves decreased in Itwari area (except increase in Pongamia) and increased in MIDC area and Azadirachta indica was found to be the most tolerant species in the environment of Nagpur (Ninave *et. al.*, 2001).

A field survey of vegetation around Thermal Power Station at Panki India was carried out by Beg *et. al.* (1990). The concentration of SO_2 and NO_X were much below the permissible limit whereas the total suspended particulate matter in ambient air was higher than permissible limit. Chlorophyll content of leaves of plants growing around thermal power plant was decreased at polluted locations. Eight out of ten plants studied in polluted area showed reduction in chlorophyll content. Based on chlorophyll destruction the order of sensitivity of various plants was *Ficus bengalensis, Sesbania sesbane, Acacia arabica, Mangifera indica, Psidium guajava, Sygygium cuminii, Ipomax corner, Dalbergia sisso* and *Eccalyptus citriodora.* Plants such as *Mangifera indica, Ficus bengalensis* and *Azardirachta indica* which exhibited visible injury at low exposure doses in experimental studies showed greater reduction in chlorophyll in field plants at polluted locations.

Effect of Air Pollutants on Human Health

Pollutants are divided into three categories depending on their nature to cause effect on different physiological processes.

Pollutants Affecting Respiratory System

Among the extremely large number of substances in the ambient air, several of them exert irritant and inflammatory effects on the respiratory organs. The main pollutants are nitrogen oxides, ozone and other photochemical oxidants, as well as sulphur dioxide and particulate matter.

Particulate matter

The deposition of particulate matter depends mainly on the breathing pattern and particle size. Larger particles are mainly deposited in the extra thoracic part of the respiratory tract (10μm) and most of the particle (5-10μm) are deposited in the proximity to the fine airways with normal nasal breathing. With mouth breathing the proportion of tracheobronchial and pulmonary deposition increases

In the lungs, particulates slow down the exchange of oxygen with carbon dioxide in the blood, causing shortness of breath and heart gets strained. People sensitive to these conditions have respiratory diseases like emphysema, bronchitis, asthma or heart problems. Particles themselves may be poisonous if inhaled, damaging remote organs like the kidneys or liver. Swallowed mucous that is laden with hazardous particulate matter may damage the stomach.

Particulate matter produce following symptoms

Respiratory symptoms: The symptoms of the upper respiratory tract include stuffy or runny nose, sinusitis, sore throat, wet cough, hay fever and burning or red eyes. Symptoms of the lower respiratory system include wheezing, dry cough, phlegm, shortness of breath (dysponea), chest discomfort and pain.

Bronchitis: Increased particulate exposure enhances the incidence of bronchitis in exposed population. Acute bronchitis and bronchiolitis may be misdiagnosed as odema, which may get further complicated in the people with myocardial damage and increased left arterial pressure. Bronchiolitis or pneumonia induced by air pollution in the presence of pre-existing heart problems might precipitate congestive heart failure and cardiovascular mortality.

Pneumoconiosis: Certain respirable dust causes group of lung diseases that lead to appreciable fibrotic changes in the lungs.

Cancer: Certain airborne particles like arsenic and its compounds, chromates, particles bearing PAHs, nickel-bearing dust, radioactive particles may act on lung tissue and cause carcinoma. These may be transported from lungs to other parts of the body, if the inhaled particles are soluble carcinogens (Parivesh, 2002).

Donaldson *et. al.* (2001) suggested that ultra fine particles could penetrate into pulmonary interstitial spaces and thus provoke inflammation. Atmospheric particulate matter in urban areas has a clear correlation with the number of daily deaths and hospitalizations as a consequence of pulmonary and cardiac disease response.

Oxides of nitrogen (NO_x)

Nitrogen dioxides (NO_2),an irritant gas is absorbed into the mucosa of the respiratory tract. Upon inhalation 80-90% of NO_2 can be absorbed, although its proportion can vary according to nasal or oral breathing. The main target in the lung tissues are at the junction of the conducting airway and the gas exchange region because NO_2 is not very soluble in aqueous surface therefore the upper airways retain only small amount of inhaled NO_2. Exposure to NO_2 could be monitored from the nitric and nitrous acids or their salts in the blood or urine (WHO, 1987a). The maximum permissible levels of exposure to nitrogen oxide are 400μg/m^3 for 1 hr and 150μg/m^3 for 24 hours according WHO guidelines (1987s).

Nitrogen dioxide exposure can cause decrement in lung function (*i.e.* increased airway resistance), increased airway responsiveness to broncho-constrictions in healthy subjects at concentration exceeding 1 ppm. Below 1 ppm level, there are evidences of change in lung volume, flow volume, characteristics of lung or airway resistance in healthy persons. It has been established that continuous exposure with as little as 0.1 ppm NO_2 over a period of one to three years, increases incidence of bronchitis, emphysema and have adverse effect on lung performance.

Nitrogen dioxide exposure may lead to chronic lung disease and variety of structural/morphological changes in lung epithelium conducting airways and air -gas exchange region. Exposure to high levels (>1.0 ppm) of NO_2 cause estuation of bronchiolar and alveolar epithelium, inflammation of epithelium and definite emphysema.

Nitrogen dioxide in large doses can result in dysfunction of host defences by causing structural alteration in ciliated cells of mucociliary escalator, in alveolar macrophages, decrease in phagocytosis, morphological and metabolic changes. The respiratory tract provide first time protective barrier against inhaled, viable and non-viable airborne agent. Breaches in defence system might increase the risk of diseases. Potential health effects of NO_2 and their mechanism is given below:

Health effect	**Mechanism**
Increased incidence of respiratory infections	Reduced efficiency of lung defenses
Increased severity of respiratory infections	Reduced efficiency of lung defenses
Respiratory symptoms	Airways injury
Reduced lung function	Airways & alveolar injury
Worsening of the clinical status of the person with asthma, chronic obstructive pulmonary disease or other chronic respiratory conditions	Airways injury

Source: Samet and Utell, 1990

Ozone and other photochemical oxidants

The primary target organ for Ozone is the lung. Ozone exposure produces cellular and structural changes, the overall effect of which is decrease in the ability of the lung to perform normal functions (Imai *et al.* 1985 and Lippmann, 1989a). Ciliated and type –1 cells are the most sensitive to ozone exposure . Proliferation of non – ciliated bronchiolar and type –2 alveolar cells occurs as a result of damage and death of ciliated and type –1 cells (Chang *et. al.*,1991).The lung airspace location where ozone exposure cause a major lesion is the centriacinar area, which includes the end of the terminal bronchioles and the first few generation of either respiratory bronchioles or alveolar ducts (Lippmann,1989a).The permissible levels of exposure should not exceed 150-200μg/m^3 for 1 hr and 100 –120 μg/m^3 for 8 hr (WHO), 1987b).

The short term exposure of O_3 has indicated a number of acute effects and the symptoms are classified below (Avol *et. al.*,1987):

» *Upper respiratory:* nasal congestion or discharge and throat irritation.

» *Lower respiratory:* substernal irritation, cough sputum productions, wheeze and chest congestion.

The non-respiratory symptoms reported are headache, fatigue and eye irritation. Ozone, like NO_2 can induce increased non specific airway sensitivity to inhalation challenge testing with bronchonconstrictive agents (HEI,1988).These effects can be produced by exposures as short as 5 minutes and the effects progress with the duration .

Asthmatics are sensitive to O_3 exposure. The O_3 may exacerbate this disease by facilitating the entry of allergens or due to inflammation. O_3 may also produce synergistic effects with other pollutants such as sulphates and nitrates (Kleinman,1990).

The long term exposure effects of O_3 are still unclear but repeated exposure could lead to chronic impairment of lung development and functions. Recent epidemiological and animal inhalation studies suggest that chronic O_3 exposure may lead to lung aging (Lippmann, 1989b).

The nasal mucosa is also affected. At high O_3 concentration effect ranged from reversible interference with pulmonary injection proliferation of type II pneumocytes and hyperplasia and metaplasia of respiratory epithelium in the nose and the permanent pulmonary fibrosis.

The toxic mode of ozone is based on the oxidation of amino acid and polyunsaturated fatty acid in cell membrane. The most sensitive cells are those with the large surface area in relation to their volume.

Extra pulmonary effects have also been observed after exposure to ozone in particular biochemical and morphological changes in erythrocytes. Whether these effects are caused by ozone itself or by reactive intermediate or are perhaps results of pulmonary effects is yet unclear. Clinical and epidemiological study in man have suggested that exposure to concentration between 160 and 340$\mu g/m^3$ may be followed by respiratory complaints, such as coughing, dry throat, chest pain and tightness of the chest.

Ozone has several toxic effects. Ozone causes eye irritation and bronchitis, aggravates chronic obstructive pulmonary disease, mutagenesis and foetoxicity. Ozone damages tissues by rapidly oxidizing thiol containing compound and unsaturated fatty acid. Symptoms of respiratory irritation and decreased forced expiratory volume (FEV) occur at exposure above 0.3 ppm, but strenuous exercise causes these effects to develop at lower ozone level.

Ozone level of 0.5 ppm increases R.B.C. fragility. Exposure to high dose of 0.6 ppm over several hours produced airways hyper reactivity unrelated to prior stabilization dose of 0.1 ppm, which did not increase airways resistance in asthmatic.

Non-pulmonary effects of ozone exposure

A number of so-called “systemic” effects have been ascribed to ozone exposure. These include change in element in the circulating blood, chromosome breakage in lymphocytes, retardation of deoxygenation of hemoglobin, alteration in the cell membrane and nuclei of myocardial cells decrease in brain secrotonin, increase of the RNA/DNA ratio in the liver, transitory increase in liver weight and alkaline phosphates, alteration of the contralateral lung when the lung were exposed unilaterally, premature aging, morphologic alterations of the parathyroid gland and increased sleeping time from sodium pento barbital.

Ozone causes lung cell injury via deterioration of primary oxidative reaction or activation of amplification system such as the products of lipids peroxidation, mediators of coagulation and for inflammation and lysosomal enzyme system.

Sulphur dioxide (SO_2)

Sulphur dioxide represents only minimal part of automotive emission, however, this pollutant may have a synergistic effect with other pollutants. Sulphur dioxide is highly soluble in the aqueous surface of the respiratory tract. It is absorbed in the nose and the upper airways where it exerts irritant effect and the minimum concentration reaches the lungs. The high concentration of SO_2 causes laryngotracheal and pulmonary oedema. Inhaled SO_2 after absorption enters into the blood and then gets biotransformed into sulphate in the liver and finally gets excreted through urine (WHO,1987c).

The maximum permissible level of SO_2 is 365ug/m^3 for 24 hrs (USEPA guidelines level, 1982b). Variation in the 24 hrs average of SO_2 has been associated with increased mortality and morbidity and reduction in lung functions.

The short-term peak concentration of SO_2 may also increase morbidity with asthma and chronic bronchitis. The long term effects of exposure to diesel engine exhaust (main source of SO_2) on the lung have been reviewed showing decrements in lung functions and increased prevalence of respiratory symptoms (Calabrese *et. al.* 1981).

Studies of normal healthy volunteers, exposed to sulphur dioxide in chambers have shown that measurable narrowing of the airways may occur after breathing the gas for 5 minutes at concentration of 4-5 ppm but the effects were not detectable at concentrations below 1 ppm. The most common acute exposure to SO_2 concentration 0.4 ppm is indication of broncho-constriction in asthmatics after exposure lasting only 5 minutes. The effects of SO_2 on airway of asthmatics are reversible with recovery occurring within one hour. Exposure at lower levels can cause increased upper respiratory symptoms such as cough, sore throat and changes in lung function. The morbidity effects are associated with long-term exposure to particulates and or sulphur dioxide. The acidic aerosols composed of particulate matter and acids cause inflammation of airways and lungs and reduce the ability of small airways to clear mucous and particles. The health morbidity indices are lung function decrement, upper and lower respiratory disease symptoms, increase in rates for cough, bronchitis and other health problems (Parivesh, 2002).

Pollutants Causing Toxic systemic effects

Carbon monoxide

Carbon monoxide is rapidly absorbed in the lungs and is taken up in the blood, where it is bound to hemoglobin (HB) with the formation of carboxyhaemoglobin (COHb), thus impairing the oxygen carring capacity of blood. The dissociation of oxyhemoglobin is also altered due to the presence of COHb in blood thereby further impairing the oxygen supply to the tissues (WHO, 1979).

The main factors conditioning the uptake of CO are concentration in the inhaled air, the endogenous production of CO, the intensity of physical effort, body size, the condition of the lungs, and barometric pressure.

The maximum permissible levels of exposure of CO is 100 mg/ m3 for <_ 15 minutes; 60mg/m^3 for >30 minutes; 30 mgm^3 for 1 hr, 10 mg/m^3 for 8 hrs (WHO, 1987d).

The main effect of CO is to decrease the oxygen transport to the tissues. The organs dependent on a large oxygen supply are most at risk, in particular the heart . central nervous system as well as the foetus. Four types of health effects are reported to be associated with CO exposure; neurobehavioral effects; cardiovascular effects; fibrinolysis effect and prenatal effect. CO leads to a decreased oxygen uptake and work capacity under minimal exercise conditions (WHO, 1979).

The classical symptoms of CO poisoning are headache and dizziness at COHb levels between 10 to 30% and severe headache, cardiovascular symptoms and malaise over about 30% (Allred *et. al.*, 1989). Above 40% there is considerable risk of coma and death (Aronow, 1981). The average COHb level in the general population is around 1.2-1.5 % (in smokers 3-4%), below 10% COHb mainly causes cardiovascular and neurobehavioral effects (Stern *et. al.*, 1988).

Decreased work capacity and neurobehavioral functions have been observed around 5%.

Low birth weight has been related to cigarette smoking during pregnancy, with the hypothesis that increased mother's COHb could have a role in this adverse effect (Heble *et. al.*, 1988 ; Ash *et. al.* 1989 and Mathai *et. al.*, 1990)

Symptoms Based on Blood Carboxy-haemoglobin Levels

Percent Hb	Medical Symptoms and Consequences
10%	No symptoms. Heavy smokers can have as much as 9% COHb
15%	Mild headache
25%	Nausea and serious headache. Fairly quick recovery after treatment with oxygen and/or fresh air
30%	Symptoms intensify. Potential for long term effects especially in the case of infants, children, the elderly, victims of heart disease and pregnant women.
45%	Unconsciousness
50%+	Symptoms Leading to Death

Symptoms Based on Carbon monoxide in Ambient Air

CO (PPM)	Exposure	Symptoms
35	8 hours	Maximum exposure limit allowed by OSHA in the workplace over an eight hour period.
200	2-3 hours	Mild headache, fatigue, nausea and dizziness
400	1-2 hours	Serious headache and other symptoms intensify. Life threatening after 3 hours
800	45 minutes	Dizziness, nausea and convulsions. Unconscious within 2 hours followed by death within 2-3 hours

Source : Parivesh, 2002

Lead

The contribution of alkyl lead additives in motor fuels accounts for the major part of all inorganic lead emissions. An estimated 80-90% of lead in ambient air is derived from the combustion of leaded petrol (Smith 1981 and Jain *et. al.* 1980). The degree of pollution from this source differs from country to country, depending upon motor vehicle density and efficiency of efforts to reduce the lead content of petrol (WHO, 1987e). About 1% of the lead in petrol is emitted unchanged as tetralkyl lead (organic lead). Concentration of tetralkyl lead amounting to more than 10% of the total lead content of ambient air has been measured in the immediate vicinity of the service station (NSIEM.1983). The WHO guidelines value for long term exposure (*e.g.* Annual average) to lead in the air is 0.5 to 1.0 $\mu g/m^3$ (WHO, 1987e).

Majority of lead in ambient air is in the form of fine particles (um). The retention rates of air borne particulate in adults range from 20% to 60%. Young children inhale proportionally higher daily air volume per unit measure (weight, body area) than adults. The children have a 2.7 fold high lung deposition rate of lead than adults on a unit body mass basis (Poenkae *et. al.*, 1993). The proportion of lead absorbed from the gastrointestinal tract is about 10% to 15% in adults and 40% in children (Frank, 1991). The absorption of lead by the lung is rapid and complete. The absorption of lead through skin also takes place but in less amount. The tetraethyl lead is metabolized (NSIWEM, 1983) in the liver and the trialkly lead is the most toxic metabolite formed

The absorbed lead is distributed among three compartments, blood, soft tissues and mineralizing tissues (bones and teeth). About 95% of the absorbed lead accumulates in the bones of adults as compared to 70% in children.

As the levels of blood lead increases, the effects become more pronounced. Lead levels of more than 50 µg/dl manifest clinically, with life threatening complications occurring at levels of 100 - 150 µg/dl. At levels < 50 µg/dl, sub-clinical effects occur. Most important of these 'low level effects' are impaired cognitive performance and behavioral changes. Intelligence Quotient (IQ) increments are associated with increasing blood lead levels, a relationship that is clear above 10 µg/dl.

Individual blood lead levels may also vary with age and season. In certain population, where there is significant exposure of children to house dust and soil contaminated with lead, there has been an increase in blood lead levels (Parivesh, 2002).

Lead also exerts an adverse effect on the endocrine system including the gonads and reproductive system (Rohn *et. al.*, 1982). It also depresses thyroid function (Tppurainen *et. al.*1988) and impairs hepatic metabolism of cortical (Saenger *et. al.*.1984). In young children, lead exposure is also associated with a decrease in the biosynthesis of 1.2.5 – dihydroxy vitamin –D, an important metabolite of vitamin D (Mahaddcy *et. al.* . 1982 and WHO,1987e).

The central nervous system is the primary target of lead toxicity in children (Davis & Svendsgaard,1987; ATSDR, 1988 and Grant & Davis,1990). Exposure to lower concentration of lead may produce neuropsychological disorders including changes in learning ability, behavior, intelligence and fine motor coordination (Harvey *et. al.*, 1984 and Fulton *et. al.* 1987). Exposure to high concentration of lead can result in an encenphalopathy, which is more frequent in childhood lead poisoning than in adult poisoning . This could be due to the crossing of lead through blood – brain barrier in children (NAS,1972).

Exposure to high concentration of Pb may lead to functional disorders of the gastrointestinal tract (Ziegler *et. al.*, 1978 and Rabinowiz *et. al.*1988),mainly of colon. Lead may also produce damage in the kidneys leading to increased urinary excretion of amino acids, glucose and phosphate (Fanconisyndrome). After long term exposure the injury may enter into a chronic stage with fibrosis and arteriosclerotic change in the kidney (Choie *et. al.*, 1980).

Pollutants Causing Potential Carcinogenic Effects

Benzene

It is a constituent of crude oil and is present in petrol in a proportion of around 5% occasionally upto 16%. The major source of benzene is emissions from motor vehicles and evaporation losses during handling, distribution and storage (WHO, 1987f and Parkinson, 1971). Benzene concentration in ambient air of residential areas generally ranges from 3-30Ugm3 (0.001-0.1ppm) depending on the traffic (WHO,1987f)

The toxic effects of benzene in human after inhalation include disorders of central nervous system (CNS), hematological and immunological effects (Sinha *et. al.*, 1969 and Niazi *et. al.*, 1989). Toxic effects are observed at very high levels (more than 3200 mg/m^3 or 1000ppm) with the appearance of neurotoxic syndrome (Takeuchi *et. al.*, 1975). Acute poisoning may lead to death due to higher exposure associated with inflammation of the respiratory tract and hemorrhage of the lungs (Pandya *et. al.*, 1975,1991). Persistent exposure to toxic levels may cause injury to the bone marrow resulting into cancer (IARC, 1982). Carcinogenic effects have been reported in workers exposed to benzene who are more likely to develop acute leukemia than general population (WHO,1987i).

Short-term exposure causes depression of the central nervous system marked by drowsiness, dizziness, headache, nausea, loss of coordination, confusion and unconsciousness. Nose and throat irritation have also been reported following short-term exposure. A feeling of excitedness or giddiness may precede the onset of other symptoms. Benzene vapors can be irritating to the eyes.

Benzene causes a serious condition where the number of circulating red blood cells (erythrocytes), white blood cells (leukocytes) and clotting cells (thrombocytes) are reduced (pancytopenia). At this stage, effects are thought to be readily reversible. However, continued exposure of benzene can result in aplastic anemia or leukemia. Benzene also damages the bone marrow, where new blood cells are produced. Studies on benzene exposure have found changes in the immune system, which are at least partially related to the changes in the blood system. Benzene may cause effects on the peripheral nerves and/or spinal cord. Symptoms include an increased incidence of headache, fatigue, difficulty in sleeping and memory loss with significant exposures. The International Agency for Research on Cancer (IARC) has concluded that there is sufficient evidence for benzene carcinogenicity to humans.

Polycyclic aromatic hydrocarbons

Polycyclic aromatic hydrocarbons are the group of chemical emitted during the incomplete combustion of fuel (Hung .1992). Exhaust of diesel engine contains lower concentration of some gaseous pollutants but higher concentration of SPM including PAH (Hampton *et. al.* 1982,

1983). Other main sources of PAHs are smoke produced on heating coal and cigarette smoking. There are several hundred PAHs the best known is benzo (a) pyrene (BAP). PAHs are absorbed in the lungs and, guts and are metabolized by mixed function oxidase system. The latter metabolites are thought to be the ultimate carcinogens (WHO, 1987 B&G).

Evidences from experimental studies exhibit that PAHs are carcinogenic (Takada *et. al.* 1990). Epidemiological studies show that coke oven workers and coal gas workers are at increased risk of lung cancer in relation to PAHs exposure (Steenland,1986). Studies of population exposed to diesel have shown definite relationship between lung cancer and smoke levels (Garshck *et. al.* 1987).

*Classification of Selected PAH's According to Carcinogencity**

Classification	PAH Compound
IARC Group 2 A	Benzo (a) pyerene
'Probably Carcinogenic	Benzo (a) anthracene
to humans'	Dibenz (ah) anthracene
IARC Group 2B	Benzo (b) fluoranthene
'Possibly carcinogenic	Benzo (k) fluoranthene
to humans'	Indeno (123 cd) pyrene
IARC Group 3	Benzo (glin) perylene
'not classifiable'	Chrysene, Coronene, Fluorene,
	Anthrace, Dibenzanthracenne

Aldehydes

Aldehydes are absorbed in the respiratory and gastrointestinal tract. They produce acute irritant effects in the eyes. Formaldehyde causes ocular and olfactory irritation (WHO, 1989). The irritation of mucous membrane and alteration in respiration, coughing, nausea and dyspea are well known symptoms produced after formaldehyde exposure (Horvathe *et. al.* 1988). Formaldehyde exposure has been associated with cancer risk mostly in occupational subjects. The sites most frequently encountered are nasal and nasopharyngeal (Olsen *et. al.* 1984 and

Source : (IARC, 1987)

Vaughan *et. al.* 1996ab), leukemia (Stroup *et. al.* 1986 and Walrath & Fraumeni 1983 1984). Human exposure to formaldehyde should be minimized not only for its probable carcinogenic effect but also for its potential for tissue damage.

Effect of Air Pollutants on Materials and Cultural Heritage

The effects of air pollution on materials and architecture have been widely recognized over the last few decades. The monuments are our heritage and the living specimen of art and architecture. Monuments, secular or religious are also tourist attractions.

Materials

Pollutants in air adversely affect materials like textile fibers, textile dyes, metals (brass, nickel, iron and steel *etc.*) rubber, paper, leather, paints *etc.*

Monuments

Monuments are made up of rocks of different types like sandstonnes, granites, basalts and marbles etc.

Major Air Pollutants That Affect Materials And National Monuments Are

Primary Pollutants

Primary air pollutants typically come from the use of fuels. The chemistry of fossil fuel combustion can be represented very simply as:

$$\text{“CH”} + O_2(g) \rightarrow CO_2(g) + H_2O(g)$$

fuel + oxygen → carbon dioxide + water

This would not be particularly detrimental to buildings. Incomplete combustion of fuel (where there is not enough oxygen during combustion) produces carbon and water. Carbon produced during incomplete combustion damages building and materials.

Source: Peter Brimblecommbe* & Cristina Sabbioni**- Air pollution effect on cultural heritage materials, EC advanced study course, Science and technology of the environment for sustainable protection of cultural heritage.

* School of environmental sciences, University of East Anglia, UK.

** Instituto di Scienze dell' Atmosfera e del Clima (ISAC), CNR, Italy.

$$\text{"CH"} + O_2(g) \rightarrow C(s) + H_2O(g)$$
$$\text{coal/oil} + \text{oxygen} \rightarrow \text{"smoke"} + \text{water}$$

In some fuels, most notably coal or fuel oil sulphur is present at high concentrations:

$$S + O_2(g) \rightarrow SO_2(g)$$

Sulphur dioxide has traditionally been an important air pollutant in cities and oxidizes to produce H_2SO_4, which damages materials and monuments.

In addition to gases the atmosphere also contains aerosol. The term aerosol indicates in general dispersion of liquid or solid particles suspended in a gas medium, which in the specific case is air. The atmospheric particles vary from 0.01 and 50 mm diameter. The larger particles fall out quite quickly and deposit on buildings. Fine particles (<lmm) in the atmosphere undergoes condensation or chemical reactions; coarse particles (>lmm), because of their high settling velocity, are rapidly deposited at the surface earth after their emission into the atmosphere. Atmospheric aerosol is expressed in term of number or mass concentration.

Secondary pollutants and air chemistry

The liquid fuels of the 20th century contain greater amount of hydrocarbons. They release low molecular weight hydrocarbons into the atmosphere as reactive trace gases. Their reactivity has changed the chemistry of the modern urban atmosphere. In the 1950's the problem of photochemical smog was unravelled in Los Angeles where, it has now become a severe problem. The hydrocarbons (here written as RCH_3) enter into reactions and are oxidized. Although here the oxidation seems to be driven by atmospheric oxygen, in reality it is mediated by the hydroxyl (OH) radical:

$$RCH_3 + 2O_2 + 2NO \rightarrow 4H_2O + RCHO + 2NO_2$$

The hydrocarbon can be oxidized to aldehyde, ketonnes or carboxylic acids.

Nitric oxide (NO) is a common pollutant from automobiles and is oxidized to nitrogen dioxide, which is a brownish gas that can absorb light and dissociate:

$$NO_2 + hv \rightarrow O + NO$$

This again forms the nitric oxide, but also gives an isolated and reactive oxygen atom that can react to form ozone:

$$O + O_2 \rightarrow O_3$$

The nitrogen oxide can become involved in further hydrocarbon oxidation and ultimately generate even more ozone. It is ozone that is responsible for photochemical smog. It is termed a secondary pollutant because, any major polluter does not emit it, and rather it is the product of the interaction of a number of pollutants in the atmosphere.

The oxidative atmosphere can produce aggressive pollutants. Perhaps the most notable is nitric acid:

$$NO_2 + OH \rightarrow HNO_3$$

Atmospheric Dispersion of Pollutants

Once the primary and secondary pollutants are in the air, they tend to be distributed within the atmosphere. Pollutants can be emitted into the atmosphere by point sources (*e.g.* a chimney), linear sources (*e.g.* a road) or aerial sources, such as an urban area. The atmospheric dispersion responsible for carrying pollutants away from their point of origin can be described on different scales: larger scale (country), local scale (urban area), micro scale (a monument or and indoor environment)

Different phenomena occur simultaneously to disperse pollutants on the various scales:

Wind transport, where the largest scales of motion are in the horizontal direction and form the basis for general circulation

Turbulence, where mixing due to eddy currents present in the atmosphere is mainly responsible for vertical transport,

Diffusion, acting particularly on the micro scale.

Mechanisms

Mechanisms by which pollutants interact with materials and monuments are:

Deposition

All gases and particulate matter emitted into the atmosphere by primary or secondary sources (formed in the atmosphere by chemical or physical processes) are deposited back to the earth's surface. Deposition occurs through two mechanisms:

- » Dry deposition, which occurs by direct reaction or adsorption at the surface
- » Wet deposition, which involve scavenging by precipitation.

Deposition occurs as a result of various phenomena and the local microclimate can reduce or increase the deposition rate. Some deposition mechanisms vary for instance with the particle diameter. Coarse particles (11-50 μm), arise mainly from the mechanical action of winds at the earth's surface, which induces emission into the atmosphere of sea salt, soil dust and biological particles from vegetation. These tend to be deposited by sedimentation because of gravitational settling. Fine particles (0.01-1μm radius), mainly originating from precursor gases, are too small in radius to sediment and are removed from the atmosphere largely by wet deposition that is scavenged by clouds, droplets, and subsequent rainout or direct scavenging by raindrops.

Reactions

Air-pollutants react chemically on cultural materials and this is usually an irreversible process. Transformation of pollutants is an oxidative process that leads to the production of acids. Therefore nitrogen or sulphur oxides attack as nitric or sulphuric acid.

Organic chemicals like formaldehyde corrode both calcareous materials and lead-containing metals. Formaldehyde in more oxidizing environment cause heavy corrosion. The oxidation leads to the formation of formic acid and under such conditions the main corrosion products were the basic lead carbonates, plumbonacrite, $Pb_{10}(CO_3)_6(OH)_6O$ and hydrocerussite, $Pb_3(CO_3)_2(OH)_2$ as well as lead formate. The carbonates are typically more stable than the formates under atmospheric conditions. It is not the concentration of formaldehyde that is important, but rather its potential to oxidize to formic acid.

There are a few materials that do not attack surfaces as the acids. Ozone that comes in from outdoors can react with the double bonds or polymers *via* oxidation. Both ozone and some nitrogen dioxides can oxidize dyes changing their colour. Sulphides can tarnish metals by creating a sulphide layer and is reducing agent.

Effect on fabric and dyes

Fabric and dyes are affected by particulate matter, acidic gases and oxidants like O_3, NO_x, PAN. Deterioration of fabric and dye by pollutant acts as additives as they accelerate deterioration caused by biological organic chemicals, sunlight, oxygen, humidity and changes in temperature.

The fading of dyes by photochemical action involved chemical reactions in which the dye molecule got activated by sunlight. The change involved both oxidation and hydrolysis of the dye. Fabrics dyed with disperse blue –27 faded to light yellow (or colorless) in the presence of low concentration of ozone even when the fabric were not exposed to sunlight. Oxidation of the dye molecule has been suspect.

Dyes vulnerable to Nitrogen oxides reddened in Chicago and Los Angeles (USA) at 470 $\mu g/m^3$ No_2 (0.25 ppm). Ozone sensitive dye faded in Los Angeles at 412 μgm^3 (0.21ppm) but not in Chicago at 10 $\mu g/m^3$ (0.005) ppm). Ozone fading take place in rural and suburban areas of Sarasota & Phoenix (118-216 $\mu g/m^3$ ozone).

Effects on building materials

Stonnes constitute one of the most widely used materials in historic monuments and buildings. The work done over recent decades has highlighted the effects of multipollutants on stonnes and mortars, by employing three different approaches:

1. Studies have been performed on monuments and historic buildings to identify the different typologies of damage affecting stonnes and mortars
2. Laboratory tests in climate simulation chambers have been set up in order to simulate the interactions occurring between, different

materials and various pollutants (such as SO_2, NO_2, O_3 and particles)

3. Field exposure tests have been carried out with the aim of measuring the synergistic reactivity to multipollutants of stonnes and mortars when exposed in different microclimatic conditions.

Effect of SO_2

The main damage effect taking place on stonnes and mortars exposed to urban multipollutants is the transformation of calcium into sulphate, due to the deposition of airborne sulphur compounds. The process occurs through two mechanisms:

1. Dry deposition of SO_2, and its transformation into sulphate at the surface of the material giving rise to the formation of gypsum ($CaSO_4.2H_2O$).
2. Deposition of SO_4 (associated with wet deposition or aerosols), which directly attack the carbonate, leading to the formation of the gypsum layer.

The former mechanism is favored by the presence of catalysts, which may derive from the atmospheric aerosol deposited on the reaction surface.

The presence of sulphites on the surface of materials, monuments and historical buildings ($CaSO_3.0.5H_2O$), indicates that SO_2 deposition is the main cause of sulphation and thus emphasize the importance of local emissions of urban pollutants.

Because of high porosity, mortars show greater reactivity to SO_2 than stonnes. Lime mortars(*i.e.* mortars prepared using lime and sand) undergo a damage typology similar to that of sandstones, with the formation of a surface black crust and an underlying layer of disaggregated sand.

Due to SO_2 deposition these materials undergo damage by:

(a) Sulphation is the primary degradation process, leading to gypsum formation;

(b) the interaction of gypsum with the hydrated mortar components (calcium aluminates and silicates) leads to the formation of two secondary products, ettringite and thaumasie. These two secondary products cause fractures and cracks in the materials, with even more damaging consequences.

On the basis of tests performed in laboratory, the formation of ettringite depends on the composition of the binder of mortars (aluminum content), while SO_2 concentration and temperature are the parameters controlling thaumasite formation. These findings . achieved with in the EC project "Environmental deterioration of ancient and modern hydraulic mortars" (EDAMM) are also of considerable relevance to the construction sector.

Effect of NO_x and O_3

Effects of NO_x and ozone in the damage of stonnes and mortars are mainly limited to the role they play in the transformation of SO_2 into sulphate.

Effect of Particulate Matter

Particulate matter on the basis of their morphology and elemental composition, the atmospheric particles embedded within the damaged layers can be classified into three main categories: carbonaceous, aluminosilicate and metallic particles. The sources emitting such particles have been identified as domestic heating system fuelled by wood, oil, distilled oil from power plants fuelled by oil or coal or as the exhaust fumes of diesel vehicles.

The interaction between carbonaceous particles and stonnes in presence of SO_2 has been investigated in a laboratory exposure system. The results show that the amount of SO_4 formed increases in the presence of carbonaceous particles and is related to their heavy metal content.

Effect of metals

Transition metals catalysts of SO_2 have also been identified in the damaged layers. After sulphur, carbon is the main element present in the damaged layers and is composed of three fractions; carbonate carbon mainly due to the material (stone and mortar), elemental carbon (EC), produced by combustion processes, and organic carbon (OC) of both biological and anthropogenic origin.

Effect on rubber

During early 1940s, it was discovered that rubber tyres stored in warehouse in Los Angeles developed serious cracks. Intensified research soon identified ozone as the causative agent, formed as a result of atmospheric reaction between sunlight (3000-4000), oxides of nitrogen and specific type organic compound *i.e.* photochemical air pollution. The problem has become so acute in Los Angeles that tyre manufacturers add a special anti ozone compound to all tyres sold in the area.

Natural rubber comprised of polymerized isoprene unit. When rubber is under tensions ozone attacks the carbon – carbon double bond (C=C), breaking the bond . The number of crack as well as the depth of the cracks in rubber under tension is related to ambient concentration of ozone.

More rapid failure of rubber insulation in atmosphere of high ozone contents is also noticed in power, transmission substance and in telephone exchange. Rubber products could be protected against ozone attack by the use of highly saturated rubber molecule the use of wax inhibitor which will bloom to the surface and the paper or plastic wrapping to protect the surface.

Effects on other materials

Paper and most especially leather suffer from attack by the sulfuric acid produced from sulfur dioxide oxidation. Dyes particularly natural dyes are sensitive to the deposition of gaseous ozone, sulfur dioxide and nitric acid.

The rate of tarnishing of silver by sulfides is much affected by the presence of other pollutants such as chlorides.

Effect on modern materials

In modern structures aluminum and stainless steel are dominant and they are typically more resistant to primary pollutants (*i.e.* sulphur and nitrogen dioxide) Ozone plays a significant role in the corrosion of aluminum. The wide use of polymers as sealants and coatings in modern architecture also pose some problems. Particles of deposited iron or polycyclic aromatic hydrocarbons on urban particulate matter can sensitize materials like polyproylene to such degradation.

To preserve historical assets pollution control is the prime requirement. Components of atmospheric pollution should be identified and controlled to ensure a sustainable preservation and conservation of the cultural heritage in an urban scenario. Industrial activities particularly those, which emit smoke and fumes, should not be carried out near monuments and commercial activities should be prohibited around historical buildings.

Chapter - 6

Plants Response to Air Pollution with Respect to Specific Biochemical Parameters

The response of plants to air pollution can be understood by analyzing the physiological and biochemical parameters of plants.

Physiological Parameters

The epidermal cells are the first to interact with the air pollutants and also lack chloroplast. Epidermal cells hence cannot remove excess sulphur through the production and release of H_2S. Damage to plants by air pollutants depends upon the amount of pollutants entering the cells and their reactivity with cellular constituents. Plants regulate the entry of gaseous molecules through stomatal movement thereby affecting the photosynthesis and transpiration. Response of plants to gaseous pollutants varies from species to species. This is due to differences in the stomatal conductance, aperture size and permeability of plasma membrane (Yunus *et al.*, 1996).

Low SO_2 concentrations normally widen the stomatal opening without affecting the diurnal functions. High SO_2 concentration induce stomatal closure by causing injury to the guard cells and by modulating the internal CO_2 concentration in the leaf due to its inhibitory effect on CO_2 fixation. Young leaves usually show less injury as compared to mature leaves. On SO_2 exposure young leaves emit more H_2S than mature leaves (Yunus *et al.*, 1996).

Uptake of pollutant gases occurs primarily through the stomata and is therefore related to stomatal conductance. Some studies have suggested a link between stomatal conductance and uptake of NO_2 and species sensitivity (Okano *et. al.,* 1989). Changes in leaf proteins and enzymes of nitrate assimilation are likely to be more important especially in determining response to high concentration of NO_x.

Photosynthesis and stomatal conductance are the early detection of potential stress and air pollution injury to plants, but photosynthesis and stomatal conductance measurements are difficult to use as biomarkers for air pollution in the field. Measuring the loss of chlorophyll from plants is an objective measure of most visible leaf injuries and directly corresponds to damage. Chlorophyll content in Mn stressed seedlings could be an effective diagnostic tool in selecting Mn tolerant cultivators of wheat (Moroni *et. al.*, 1991). Chlorophyll content in an ozone sensitive clone compared to an ozone resistant clone of *Trifolium repens* indicated specific growth reductions caused by atmospheric ozone during the summer. (Heagle *et. al.*, 1994).

The stomatal opening responses to SO_2 may substantially increase transpiration of field crop (Rennerberg and Herschbach, 1996). Wide stomatal apertures led to high rate of transpiration, which exceeded the rate at which water could be replaced by vascular system (Heath, 1988). The brown necrotic areas that developed were similar to those caused by SO_2 in many plants thus indicating that some of acute injury due to SO_2 could in fact be the result of excessive transpiration. The effect of SO_2 on net photosynthesis varies greatly in different species of a plant. SO_2 concentrations above 0.5 ppm are strongly inhibitory to photosynthesis and stomatal closure, which in turn causes depression in net photosynthesis and transpiration (Osnubi and Davies, 1980).

Biochemical Parameters

Plant Pigments

Structural changes in pigments due to air pollution

A chlorophyll molecule is a typical prophyri derivative possessing a cyclic tetrapyrolic structure in which one pyrrole ring is partially reduced. The tetrapyrrolic nucleus contains a non-ionic magnesium atom held by two covalent and two coordinate bonds. In addition to four pyrrole rings, a fifth isocyclic ring is also present. Both acid side chains are esterified, one as methyl ester and other as phytol ester. The presence of a long phytol tail along with the flat porphyrin head gives the molecule an appearance analogous to that of common spatula. In chlorophyll b, the methyl group at position 3 of second pyrole ring of chlorophyll a is replaced by a formyl group. In its natural state, chlorophyll is bound to protein to form a lipoprotein complex.

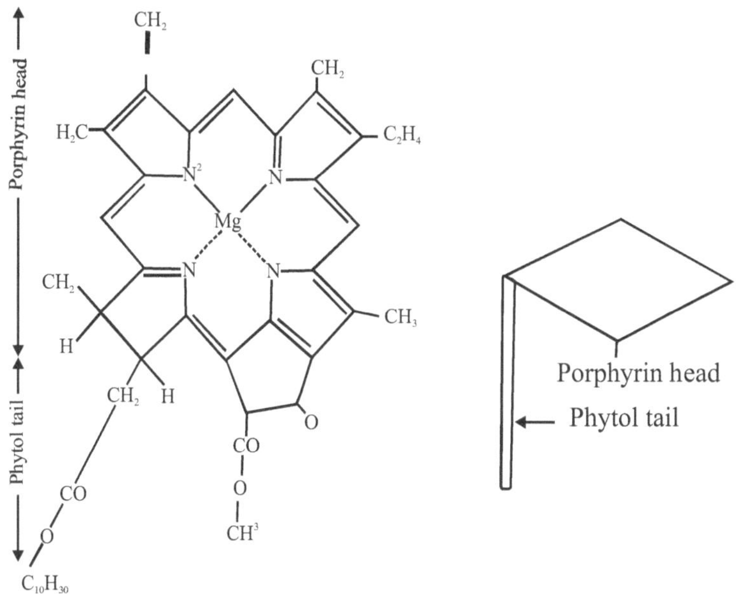

Rao and LeBlanc (1965) observed the destruction of chlorophyll molecules in lichens following exposure to large doses of gaseous SO_2. They described that in this process Mg^{2+} of the chlorophyll molecule is replaced by two atoms of hydrogen forming phaeophytin, thus changing the light spectrum characteristics of the chlorophyll molecules. Reports of decreased chlorophyll in plants fumigated with SO_2 are often associated with degradation of chlorophyll a, however, instances of simultaneous destruction of chlorophyll a are also known (Singh *et al,* 1998, Khan *et al*, 1990, Singh *et al* 1990).

Degradation of photosynthetic pigments has widely been used as an indicator of pollution. Chlorophyll, the green pigment is one of the main complex, which influences photosynthesis. The degradation of chlorophyll pigment may be attributed to the phytotoxic action of SO_2. Breakdown of chlorophyll to phaeophytin under the effect of SO_2 has been experimentally proved. All the plant species, whether studied under transfer experiment, laboratory experiment or field survey, showed decreasing amount of chlorophyll. Chlorophyll damage hampers photosynthesis and subsequently the plant productivity. Reduction in net photosynthesis has been observed in pollutant-exposed plants. Darall (1986) reported that SO_2 might reduce plant growth even at concentration, which is not enough to cause visible injury, mainly through its adverse effects on photosynthesis.

The solubility of SO_2 is very high in aqueous medium and it is because of these characteristics SO_2 gets dissolved in the water matrix available on the cell surface and subsequently infuses in cell. The infusion of H^+, HSO^{3-}, SO_3^{2-}, and SO_4^{2-} ions brings down the pH of the cell in the acidic range, which is not suitable for the carbon fixation. Out of the four ions, sulphite ions are more destructive than other sulfurous compounds. Variation of chlorophyll with pollution level and with the season is *described in chapter 5 and chapter 12.*

Decrease in chlorophyll under stress (pollution or temperature stress) may be attributed to either its degradation or to reduced biosynthesis. H^+, HSO^{3-}, SO_3^{2-} and SO_4^{2-} ions, which are generated by the dissolution of SO_2 in the cytoplasm are incorporated into the thylakoid membranes and induce swelling (Ziegler, 1977, Welburn *et al.*, 1972) or disintegrate membrane.

The adverse effects of SO_2 on chlorophyll pigments leading to reduced productivity may be considered fewer than two cellular pH conditions. At pH 2.2 to 3.5, the free H ions generated in the cell from the splitting of H_2SO_3 into SO_3^{2-} and H^+, displace the Mg^{2+} from the chlorophyll molecule to degrade to latter into Phaeophytin molecule a non-photosynthetic brown pigment (Rao and Lee Blanc, 1966). At pH above 3.5, SO_2 affects thylakoid membrane of chloroplast by causing oxidation of carotenoids through generation of O^{2-} from HSO_3 (Pieser and Yang 1978). The unprotected chlorophyll then is oxidized and lost. Free O^{2-} also increases level of H_2O_2 in the presence of Superoxide Dismutase (SOD) leading to the oxidation of chlorophyll molecules. SO_2 is also considered to reduce Chlorophyllide synthesis through its effect on ascorbic acid (Keller and Schwager 1977).

Like chlorophyll carotenoids occur in their natural state as protein complexes. Chlorophyll and carotenoids may be attached to the same portion forming a complex known as photosynthin. The ubiquitous presence of carotenoids in the photosynthetic tissue suggests a fundamental role in the photosynthesis process. Major role of carotenoids are:

» Carotenoids protect against photodynamic destruction catalyzed by chlorophyll
» These pigments absorb and transfer light energy to chlorophyll a
» Carotenoids are the scavengers of free radicals and represent the anti oxidative system of plants.

Carotenoids belong to a large group of compounds called terpenoids. These compounds produce red orange, yellow, cream and brown color in plants. They are further divided on the basis of presence of and absence of oxygen into:

Carotenes, which have formula $C_{40}H_{56}$, contain only C and H.

Xanthophylls contain oxygen in addition to H and C *e.g.* common xanthophyll of leaves is lutein ($C_{40}H_{56}O_2$).

Carotenoid content of plants varies species to species. Plants with high baseline value of carotenoid show that they are better equipped with anti oxidative system. Pollution produces oxidative stress in plants. It is observed that under the pollution stress carotenoid content of some plants increases (refer chapter 12). It shows that under pollution stress defense system of plants gets activated to reduce the burden of pollutants.

Protein

Protein is the basic constituent of the cell. These are complex substances of high molecular weight ranging up to several millions and contain nitrogen in addition to carbon, hydrogen and oxygen. Sometimes elements like phosphorus, sulphur, iron, zinc and iodine may also be present. However elementary composition of most proteins is very similar (approximate percentages are C=50-55, H=6-8, O=20-23, N=15-18 and S= 0-4). Proteins are made up of several nitrogen containing organic molecules called amino acids *i.e.* proteins are polymerized forms of organic molecules called amino acids. Thus amino acid is the basic unit of protein. Protein dissociates to form amino acids and the energy produced is utilized for routine metabolic activities.

Proteins are the most important constituent of plant cells both, from structural as well as functional point of view. Functionally, give rise to enzymes, which are responsible for regulating the cellular processes. Many proteins carry out enzymatic activities and are vital for the rapid rate of biochemical reactions in the cell. Two important functions of proteins are

(a) These are hydrogen ion buffers

(b) These are structural components of cell

Nandi *et. al.,* 1990 observed decreased protein content in plants fumigated with SO_2. This decrease might be due to the breakdown of proteins or reduced de novo synthesis. In the latter phenomena decrease in protein content is not followed by increase in amino acid content. Reduction in protein availability by either of the above mentioned ways affects various metabolic functions and also limits the availability of proteins, a structural component of new membranes and H+ ions buffers (Khan and Malhotra, 1983)

Some plants show an increase in protein content in polluted areas. The increase in protein content may reflect enhanced de novo protein synthesis as the plants attempt to overcome pollutant injury.

Peroxidase Enzyme

Peroxidase enzyme belongs to the class of oxido-reductase, occurs in a wide variety of trees and shrubs and has been extensively studied in attempts to develop a method for early identification of chronic injury from air pollution. The enzyme decomposes hydrogen peroxide with a current oxidation of various substrates. Plant cells showed increased peroxidase activity under a variety of stresses such as mechanical injury, attack of pathogen or an influence of environmental pollution (Nandi *et. al.* 1990). Air pollutants *e.g.* O_3, NO_x and SO_2 produce toxic oxygen free radicals. (Molecules containing unpaired electron). The generation of radical species can occur through oxidation-reduction or ionizing reactions. Dioxygen reduction commonly occurs in both biotic and abiotic systems and intermediate of this reduction (oxyradicals), are the superoxide radicals (O^-_2), hydrogen per oxide (H_2O_2) and the very reactive hydroxyl radical (OH^-).

Plant cells exposed to SO_2, peroxidase activity increases when sulphite is oxidized to sulphate in the presence of hydrogen peroxide and peroxidase. This increase in peroxidase activity varies species to species, season and concentration of pollutant. The changes in the peroxidase activity are accompanied by opposite changes in the catalase activity. The increase in peroxidase activity was accompanied by an increased accumulation of sulphur in leaves, a decline in CO_2 assimilation and a reduced content of ascorbic acid.

Peroxidase may serve as a good indicator of pollution caused by O_3, NO_x, H_2O_2, and SO_2. A change in peroxidase activity was observed prior to the visible injury. Ultra structural studies have shown that peroxidase is confined to cell walls, endoplasmic reticulum and golgi apparatus. In sensitive trees, enzyme activity is higher in mesophyll tissue while in resistant trees the activity is higher in the epidermis, hypodermis and in the resin canals. The activity was drastically high in the cells of the hypodermis and endodermis of trees resistant to SO_2 (Yunus *et al*, 1996).

Free Radical Injury To Plants

Air pollutants like O_3, NO_x, H_2O_2 and SO_2 produce toxic free radicals. These are generated by oxidation, reduction or ionizing radiations. Reduction of molecular oxygen produces superoxide, hydrogen and hydroxyl radicals. Dioxygen reduction commonly occurs in biotic and abiotic systems and the intermediate of this reduction (oxyradicals) are the superoxide (O^{2-}), hydrogen peroxide (H_2O_2) and the very reactive hydroxyl radical (OH.). The extremely reactive hydroxyl radical oxidizes organic molecules including bimolecules, which results in enzyme breakdown, membrane damage (lipid peroxidation) and DNA alterations. The primary defense against the oxygen compounds superoxide and hydrogen peroxide involves the enzymes super oxide dismutase (SOD) peroxidase (P_x) and catalase (Richardson *et. al.*, 1989).

Reduction of Molecular Oxygen

A free radical is defined as any group of atom or molecule in a particular state with one unpaired electron occupying the outer most orbit. The one electron oxidation or reduction of compounds results in the production of either cation or anion radicals respectively (Thomas and Aust, 1986). The univalent pathway for the reduction of molecular oxygen results in the generation of super oxide ion radical the reduction of (O^{2-}), H_2O_2 and OH.. Intermediates are shown in the figure 1. In addition to direct oxidative degradation of lipids, destruction of extra cellular and intercellular proteins including enzyme inactivation or DNA damage, indirect toxic effects like mutagenecity and carcinogenecity have been related to oxygen free radical formation (Kappus, 1987)

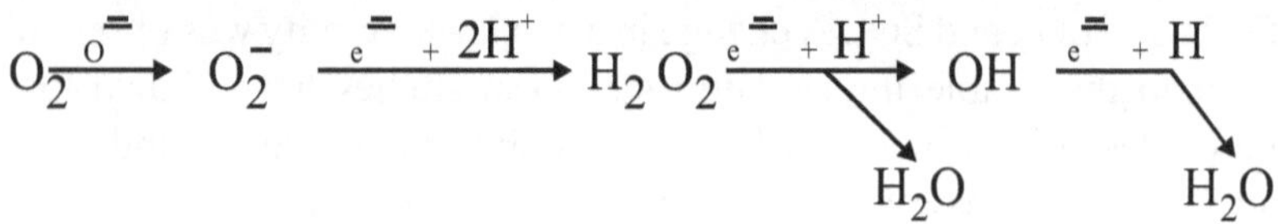

Figure 1: Reduction of molecular oxygen (from Del Maestro, 1980)

Aerobic cells have evolved variety of enzymatic mechanisms against the toxic effects of the intermediates of oxygen reduction (oxyradicals) such as SOD, P_x catalase shown in figure 2. In addition non-enzymatic defense include compounds such as glutathione, vitamin E, carotenoids and ascorbic acid (Di Giulio and Richardson 1987).

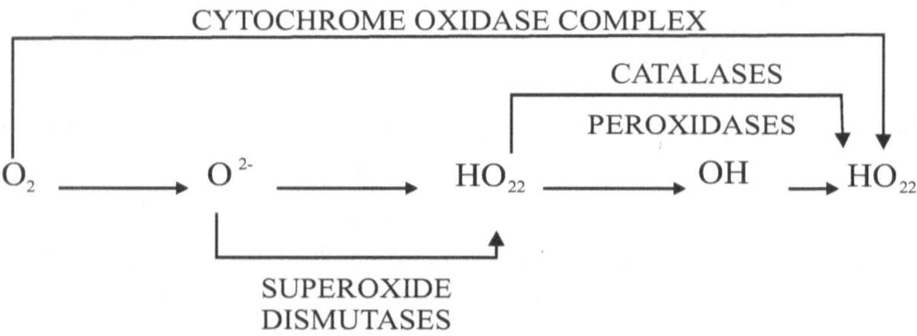

Figure 2: Defense mechanism against free radicals (from Del Maestro, 1980)

Gaseous pollutants associated with acid precipitation (SO_2, NO_x and O_3) are the principal source of free radical mediated toxicity in trees. A model of how SO_2 affects chloroplast and cytoplasm was given by Schulz in 1986 is shown in figure 3. Once SO_2 enters the leaf it is converted to bisulphate and sulphite. Bisulphite is photo-oxidized to less toxic SO_4 and the super oxide radical (O^{2-}). The super oxide radical can be dismutated by SOD to H_2O_2 and O^{2-}. Px and catalase convert

H_2O_2 to H_2O and O_2. Toxicologically, O^{2-} and H_2O_2 are considered particularly important precursors for the highly reactive and destructive OH radical. Cell injury can be severe if even a small portion of the superoxide radical is converted to OH.

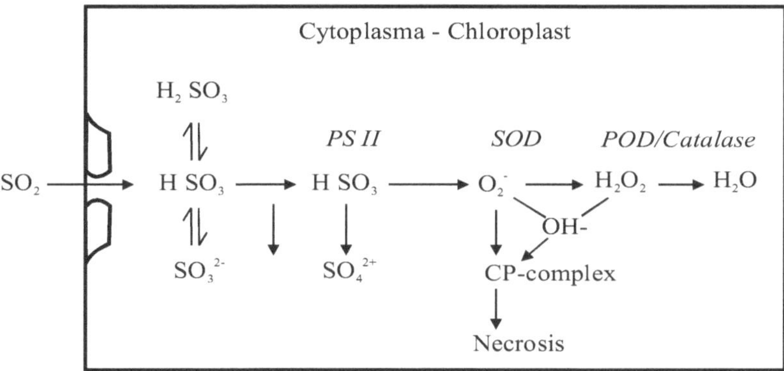

Figure 3: Effect of SO_2 on chloroplast and cytoplasm as proposed by Schulz

Chapter - 7

METEOROLOGY AND AIR POLLUTION

Atmosphere serves as the medium through which pollutants are transported and dispersed. While being transported, the pollutants undergo chemical reactions, which may be lethal or scavenged by chemical or physical process/method. But it all exclusively depends upon the meteorological conditions (Lee *et.al*, 1994). Historic air pollution episodes reveal that atmospheric conditions influence the pollutant concentrations as well as their toxicity (Goldsmith & Friberg, 1976; Wichmann *et.al*, 1989 and Pope *et.al*, 1992). However, the location, duration of release, height as well as the pollutant concentrations are also important, which effect the secondary pollutant transformation under the influence of meteorological conditions, but the atmospheric behavior is independent of their sources (Holzworth, 1974).

Meteorological variables and auto exhaust have shown significant correlation with atmospheric chemical process (Sjoedin *et. al,* 1994 and Alary *et. al,* 1994). The chemical processes occurring in the atmosphere involve oxidation of hydrocarbon, sulphurdioxide (SO_2) and nitrogen oxide (NO) to form oxygenated product such as aldehyde, nitrogen dioxide (NO_2) sulphuric acid (H_2SO_4). The oxygenated species become the secondary product formed in the atmosphere from primary emissions of anthropogenic or natural sources.

Urban atmosphere is characterized as complex mixture of hydrocarbons, peroxyacetyl nitrate (PAN), polycyclic aromatic hydrocarbons (PAH), oxides of sulphur and nitrogen (Lonneman *et. al*, 1978). The concentration profiles are the results of the combination of atmospheric chemical and meteorological process (Solane and Tesche, 1991 and Wameck, 1988).

Solar radiation influences the atmospheric chemical processes. Free radicals are formed by photodissociation of molecules and these are highly reactive unstable intermediates. Although, solar radiation plays crucial role in the generation of free radical yet water vapour and temperature also influence the particular chemical pathways.

Photochemical dissociation of NO_2 takes place due to absorption of solar radiation while ozone is formed due to combination of free radical oxygen and molecular oxygen (O_2). The Nitrogen oxide (NO) is oxidized by ozone (O_3) to form nitrogen dioxide (NO_2) and molecular oxygen (O_2).

The photochemical oxidation leads to formation of toxic compounds like peroxyacetyl nitrate (PAN). It is formed when peroxyacetyl free radical reacts with nitrogen dioxide (NO_2). The free radicals are produced from degradation of hydrocarbons. Hydroxyl (OH) and H_2O oxidize SO_2 to reactive intermediates such as HSO_3 and SO_3. These intermediates combine with water vapour in the atmosphere to form sulphuric acid aerosols. These processes depend upon the atmospheric conditions (Calvert *et. al*, 1978). The homogeneous oxidation of SO_2 by free radicals leads to the formation of photochemical smog. Smog problem is accelerated with the presence of automobile exhaust in the environment hence as such the effects of this problem are more in urban area than in rural one (Interim Report, 1996). Smog, is the result of smoke and fog and the humidity plays crucial role in the condensation of smog. Combined activity of meteorological conditions like high pressure, inversion of temperature, stable atmosphere and strong solar radiations also help in the formation of smog. The smog results in impairment of vision, eye irritation, and short time acute respiratory trouble (Buma, 1960).

In various studies it has been shown that monsoon is the best season for removal of air pollution in comparison to that of summer and winter. Winter is the worst among other seasons because of adverse meteorological conditions (Firket, 1936: Schrenk *et. al*, 1949; Wilkins *et al*, 1954; Heylin, 1985 and Vardrajan, 1995). Historic air pollution episodes revealed that meteorological conditions were involved in the smog, formation.

Major air pollution hazards due to meteorological conditions

In heavily industrialized Meuse valley in eastern Belgium an intense fog occupied the entire valley in December 1930 (Firket, 1936). More than 100 persons suffered from respiratory problems and 63 persons died after few hours of sickness. Fog dissipated in a week time and the respiratory problems of the people living in vicinity improved afterwards. Fog was generated under anticyclonic conditions when temperature inversion extended to a height of about 90 m. The fog was cooled by radiation from the top and warmed by contact with ground, which resulted in the development of stable environment. However, there were no measurement of pollutants during the episodes but enquiry indicated that sulphur dioxide was probably oxidized to sulphuric acid, which affected adversely the human population living in the vicinity.

In Donora Pennsylvania during October 25 to 31, 1948, severe atmospheric pollution occurred (Schrenk *et al*, 1949). Out of twenty persons seventeen died within 14 hours on October 30. During this period cold air accumulated in the bottom of river valley and fog was formed. The top of the fog layer had high albedo and reflected solar radiations. During night further cooling stabilized the layer and the wind speed decreased to less than the 3.1 m/sec (11.16 Km//hr). The pollutant remained in the valley and did not transport far from the site due to less wind speed. These conditions indicated extreme vertical stability of atmosphere. In the vicinity of Donora, sources of emission of sulphur dioxide, particulate matters and carbon monoxide were found to be from industries.

Another fog episode occurred during December 5 to 9, 1952 in the London (Wilkins, 1954 and Wilkins Jr., 1954)). Fog appeared as the anticyclones approached from northwest. There was no wind and the atmosphere became almost stationary. Temperature remained near the freezing point. The depth of fog was variable and it was 100 m or less. The visibility was restricted for 4 days to less than 20 m over an area of 40 km. From the commencement of the fog & low visibility several people experienced difficulty in breathing and daily deaths reached to peak on December 8 and 9. The problem decreased with the decipiation of fog and the deaths were correlated with the synergetic effect of fine particles and sulphur dioxide. It is believed sulphuric acid mist was formed from oxidation of sulphur dioxide but its amount was speculative since no measurements were made.

On December 2, 1984 when people were sleeping in their houses in the night (Heylin, 1985 and Vardrajan *et al*, 1995), methyl isocyanide (MIC) was accidentally released from storage tank of Union Carbide India plant at the Bhopal. After midnight, gas spread like fog over thickly populated area in the south and east of the plant. Moreover 2000 people died and 1000 people or more suffered. Medical examination records indicated eye irritation, impaired vision and damaged lungs in exposed persons (Sriramachari & Chandra. 1997). The atmosphere at that time of M~IC release was relatively stable and winds were quiet light and temperature was also low.

Meteorology and particulate dispersion

Micrometeorological parameters such as wind speed, wind direction, temperature, atmospheric pressure and relative humidity have great effect on the movement of particulate matter as well as the concentration level at any particular area. Sedimentation, diffusion, turbulence, washout, occult depositions are the processes, which remove particulate matter from air. The attractive and reactive properties of some particles cause their agglomeration into larger particles, which may then fall to earth *via* sedimentation. Agglomeration is initially driven by the forces, which cause particles to meet. These range from Brownian movement at the nanometer scale, to wind and convective air currents above the micron size range. Sulphate and nitrate particles, in particular, can readily absorb moisture and expand (hygroscopic growth); these can then fall as occult deposition, or if humidity is high enough, as raindrops. Particles can also be 'washed' out of the atmosphere from impacts with falling precipitation air (Beckett *et. al.*, 1998).

Deposition of particles from the atmosphere occurs by the action of wind turbulence on a large scale when wind speed falls in the laminar sub layer due to the presence of topographical or meteorological obstruction; *e.g.* a forest/bare ground boundary in the former case and a cold weather front in the latter (Pye, 1987). Larger, heavier particles will fall to earth much quicker than finer particles due to their higher settling velocities. This is the reason for the higher rates of deposition closer to particle emission sources (Boubel *et. al..*, 1994). The residence time of sub-micron particles in the atmosphere is, therefore, much longer than for larger particles. The Quality of Urban Air Review Group's report (QUARG, 1996) stated that the lower deposition rate results in a 'background' concentration of particles from a variety of widespread sources, covering large areas, and even permeating the indoor environment. The implications of indoor air pollution are relatively new and were discussed by Crump (1996) and Zummo and Karol (1996). In addition, use of plant surfaces to reduce particle concentration in indoor air was discussed by Lohr and Peerson-Mims (1996). Such background levels of particles were measured by Monn *et. al..* (1995), and found that PM_{10} concentrations at a number of rural areas (Swiss) were very similar due to the mixing of ambient particulates from a variety of disparate sources. They also found that in these areas it was the finer fraction of particles between 0.3 and 0.7 μm diameter that contributed most to the PM_{10} total mass. Janssen *et. al..* (1997) found that the larger diameter component of urban PM_{10} consisted mainly of locally deposited carbon and road dust. These were, therefore, excluded from the finer, homogeneous, background particle mass. This level spread of pollutants is special for particles since gases such as SO_2 and NO_x are removed from the atmosphere relatively quickly due to reaction with other compounds (Score, 1973).

Study by Shen *et. al.* (2002) revealed that SPM and settleable dust were maximum during summer season followed by winter and minimum during rainy season (Beckett *et. al.*, 1998). Particles exist in the atmosphere in many forms, from sub-micron aerosols to clearly visible grains of dust and sand. The effects of such particles on human health have been made readily apparent in our history. The London's smog of the early 1950s, which had a significant particulate content, claimed the lives of many people and caused acute illness in many more, culminating in the passing of the first Clean Air Act of 1956 (Department of the Environment (DOE, 1995a). The capture of particles by solid objects, both synthetic and organic, occurs, by a variety of impaction processes. Particles in an air stream are most readily impacted onto moist, rough, or electrically charged surfaces (Pye, 1987).

McPherson *et. al..* (1994) estimated that in a model of PM_{10} interception by urban trees, 50% of deposited particles were re-suspended by 'normal' meteorological conditions. Electrostatic and radiometric forces are rarely important in the attraction of particles in the natural environment. Retention of impacted particles is increased if the particles are of a relatively high mass or if the object is made sticky (Gregory, 1973). However, at sufficiently high wind speed impaction can be reduced to almost zero due to increased 'bounce off'. Once impacted, it requires strong forces to dislodge the majority of particle from a surface (Lyons and Scott, 1990). Direct interception occurs when particles are deposited by an air stream or gravity forcing it onto the surface, and inertial impaction occurs when the momentum of a particle carries it through the air stream, which curves around the surface. Particles will also be deposited downwind of the edge of a suitable object, such as a leaf or stem, due to the action of turbulent eddy currents (Gregory, 1973; Chamberlain, 1975).

The residence time of particles will also be greatly increased during extended periods of warm weather and low humidity because the-washout process and hygroscopic growth will operate less effectively (Fenn and Byterowicz, 1997).

Air quality monitoring was conducted to quantify the spatial and temporal variations in the concentrations of Total Suspended Particulate Matter (TSPM) and dust fall rate around a cement factory situated in a dry tropical area. Particulate matter showed a linear spatial decrease as one moves away from the source. Concentrations of TSPM often exceeded permissible limits up to 2 km SE of the source. Dust fall rate was quite significant up to 5 km SE. Dust fall rate was maximum during winter, while TSPM levels were maximum during summer (Agrawal and Najma, 1997).

SEASONAL VARIATION AIR POLLUTANTS

*Monsoon**

The pollutants get washed during monsoon and settle down on the earth due to heavy rainfall and thunder and therefore, the quality of ambient air is good in this season. The concentration of gases and particulate matter was remarkably decreased in this season as compared to other seasons. This is significant as it establishes the correlation between the meteorological factors and pollutant concentrations. The high relative humidity, moderate temperature and heavy rains result in

---

**Source:* Thesis entitled Air pollution and related health effects due to auto exhaust and other sources in Lucknow city, By Dr. Sanjay Pradhan, University of Lucknow, Lucknow)

the decrease of concentration of particulate matter, oxides of nitrogen (NO_x), sulphur dioxide (SO_2) and formaldehyde (HCHO). This also indicates that there will be less impact on the health, vegetation and materials during monsoon due to air pollution.

Summer

Summer can be considered, as good season but not like the monsoon. In this season temperature shoots very sharply as a result urban areas become warmer during the day and store sufficient heat which is reradiated in the night, consequently urban atmosphere becomes more warmer as compared to its surrounding rural area. The light wind blow through night as a result of this auto exhaust both solid and gaseous pollutants further gets warmed in dusty strong heat waves. The urban plumes along with the pollutants move in upward direction. The conviction in planatory boundary layer provides natural dispersive force, which dilutes the ground level concentrations of both the solid and gaseous pollutants. The convictive currents -carried the pollutants upwards and mix upto altitude of 1 km or more. The prevailing winds at higher altitudes disperse the pollutants. The high wind speed eventually dilutes the pollutants in the atmosphere ultimately pollutants dispersing in the wind range. The relative humidity is found relatively low as compared to other seasons.

Winter

The magnitude of air pollution in the winter season is found very high and considered, as the worst and therefore, higher health hazards in this season cannot be ignored. The temperature comes down sharply by the end of November and starts increasing from the end of February. Ground level mixing height was decreased during winter, starting from mid November and continuous till the end of February. The effect is more prominent between 4.30 P.M. to 8.30 A.M. next day. In view of this ground level inversion and fog formation is common in this part of season. The problem of fog formation is acute on the busy roads and in the peak hours having high density of vehicular movement. The emission from these vehicles does not go up and therefore higher concentration of SPM, and NO_x were observed during these periods. As a result of this the photochemical smog formation occurs in the peak hour and effect was found more prominent in the evening. Due to this visibility becomes poor in the evening hours. The situation becomes better in the morning (after 8.30 A.M.) when mixing height rises to more than the 500 m. This shows the formation of surface based inversion, which starts in the evening, and it grows in time in term of both the magnitude and the thickness of inversion. The inversion decays slowly after sun rise. Relative humidity is very high early in the morning. This keeps the humidity content of the surface layer very high.

The temperature of the lower atmosphere suddenly enhances by 0.5°C. This process decreases the steep inversion in the Planatory Boundary Layer (PBL). In the winter pressure is maximum hence anticyclones approach. The convictive currents are very light and rise with urban plume upto few hundred meters to an altitude of 1 km. The prevailing winds are very light hence dispersion of pollutant is very slow. The dissipation of rising layer takes maximum time in the winter due to low solar heating of the ground and the formation of inversions starts early in the evening due to strong solar radiative process. Coupling of atmospheric eddies and surface based inversion results in downward move of air pollutants, which occupy large volume of air in the breathing zone or near to the surface layer. This ultimately leads to greater degree of exposure to pollutants to the general population and results in the reduction of visibility, eye irritation, suffocation and respiratory problem in the peak hours.

It is noticed that the low vegetation, increased industrial, automobile activity and domestic air pollution results in an increase of atmospheric temperature and relative humidity in urban area. If current predictions are borne out, a global warming associated with high concentration of atmospheric pollution is eminent. Global warming is likely to change the climate of the world during the next century. The threat of climate change is sufficiently great and requires immediate action.

Chapter - 8

HEAVY METALS IN URBAN ENVIRONMENT

Metals

Metals are ubiquitous in the modern industrialized environment. Heavy metals like cadmium, iron, molybdenum, lead, nickel, tin and zinc are present in the environment in trace concentrations. These metals are universally present in soil, water, air and biota (Freedman, 1995). Global anthropogenic heavy metal inputs to ecosystem through air, water and soil have increased substantially world wide over last century (Nriagu, 1996).

Other substantial contamination sources of heavy metals are fumes and ashes formed during combustion of fossil fuel and of municipal and industrial wastes (Salmons, 1993). Heavy metal contamination and the problems that they poses to the biota have been well documented (Raskin and Ensley, 2000 and Meagher, 2000).

Recent studies have shown that urban environment is contaminated by heavy metals from vehicular sources. Many trace metals are present in leaded and unleaded petrol, diesel oil, antiwear substances added to lubricants, brake pads and tyres and are emitted by vehicle exhaust pipes (Monacci and Bargagli, 1997). In 1988, London Scientific Services (LSS) presented results of base line metal in dust sampling survey in greater London. Industrial areas and areas with high traffic density have the highest concentration of cadmium, copper, iron, lead and zinc (Schwar *et. al.*, 1988).

Motor exhaust was the most important source of metals in air (Colombo *et. al.*, 1999). Trace metal concentration followed the behavior of total suspended particles with higher concentration during the day and lower at residential site. In general spatial and temporal variations prevailed over diurnal differences. Spatial differences were clearly most significant for Pb, which represented higher values at the traffic site reflecting the importance of motor exhaust inputs. In contrast diurnal differences were more important for Mn due to increased dust suspension during day hours.

Metals from air wash to the soil and pollute the soil. This is of great concern for the environment (Ferguson and Kasamas, 1999). The soil, the common rooting medium, is the primary source of metals in plants. In general, as the trace element content of soil increases, that amount available to the plant also increases, although there are other factors (such as soil pH, level of organic matter, texture *etc.*) that will determine what portion of the trace element content of soil will be available for root absorption (De Temmerman *et. al.*, 1984). Dissolved organic compounds in the soil increas the availability of metals to the plants (Antoniadis and Alloway, 2002).

Anthropogenic enrichment of soil by heavy metals is mostly by metal oxides, often combined with the emission of high concentrations of sulphur dioxide enhancing the availability of metals to plants with consequences for plant survival and selection of metal tolerant species (Wilfried, 1999).

The activity of soil and water borne bacteria and fungi causes some metals to bind with organic substances, which in many cases substantially increase their toxicity. The increased number of toxic metals compounds in the soil often manifests by an intrusion into the food chain and a subsequent increased risk to human health (Plette *et. al.*, 1993)

Addition of 0.1% (v/w) of diesel to an artificial agro ecosystem, there occurred a decline in microbial population within 7 days, which later increased gradually and stabilized after 63 days (Bhaduria Seema, 1999).

Liming the soil reduced the heavy metal content in plant tissue, except for iron. Chemical speciation plays an influential role in the solubility and potential bioavailability of heavy metals in the soil. Metals in the exchangeable and carbonate-bound fractions are considered readily and potentially bio-available (Wong *et. al.*, 2001).

Soils in certain areas of El Paso, Texas have sites that have lead and copper concentrations as high as 5067 mg/kg and 4955 mg/kg respectively. These are the concentrations well above the permissible limit (Gardea Torresday *et. al.*, 1996).

Ray (1990) found significant bioaccumulation of heavy metals in vegetables and indigenous weeds in the polluted field. Maximum accumulation of metals was found in the edible parts followed by the non-edible parts. The least accumulation was found in the leaves and shoots (Barman and Lal, 1994).

From the soil potential toxic elements are transported to plants (Wenzel and Jockwer, 1999). The mineral content of plant reflects many different aspects of their environments and foliage analysis has been used as valid indicator of the air quality (Aksoy and Ozturk, 1997). The uptake of heavy metals *via* plant roots is generally the most important pathway of absorption. The extent of heavy metal uptake depends upon the ionic potential of the metal; this is defined as the ratio of the change to the ionic radius. The translocation process of heavy metals in crop parts is influenced by mechanisms including anatomical, biochemical and physiological (Salt *et. al.*, 1995).

Sensitive plant species effectively accumulate metals in their tissue, which may have some relevance with chlorophyll reduction. Ficus *bengalensis, Psidium gujava* and *Sygygium cuminii* were the plants, which trapped higher amount of metals in leaves around thermal power station (Beg *et. al.*, 1990). Some heavy metals in higher doses may cause metabolic disorders and growth inhibition for most of plant species (Clarie *et. al.*, 2001). Plants have the ability to grow in sites where soils contain greater than usual amounts of heavy metals or other toxic compounds (Raskin and Ensiley, 2000).

The amount of heavy metal uptake by plants was significantly affected by the dose of the metal in the media and by tissue of the plant. Peralta *et. al.* (2001) suggested that metal uptake by Alfaalfa was in the order of Zn > Cu > Cd > Ni > Cr. Soils moderately contaminated with Cd, Cr, Cu and Ni did not cause adverse effects in plants.

A common reaction of plants to metal stress is the induction of Peroxidase. Activation of Oxygen has been hypothesized as a general reaction during several kinds of stress including heavy metals (Singh *et. al.* 1997). Rise in Peroxidase level during metal is therefore a protection mechanism for the preservation of biological membranes.

Heavy metal ions such as Cu, Zn, Mn, Fe *etc.* are essential micronutrients for plant metabolism but when present in excess, these and non essential metals such as Cd and Pb can become extremely toxic (EEA, 1998).Because of their capacity to act as efficient interceptors and accumulators of chemicals, plants are widely used as passive biomonitors in urban and rural environment (Monaci and Bargagli, 1997).

Presence of heavy metals in soil and accumulation by herbs, shrubs and trees in the chromite mines was studied by Samantry *et. al.* (1999). There was significant correlation between Cr and Fe in the soil where a tree species occurred. Trees showed significant correlation between stem, leaf and soil for Cr and Fe.

Metal concentrations (Cu, Pb and Fe in Quercus ilex) were significantly higher in leaves from roadside sites than in the leaves collected from town, which in turn were significantly higher than concentrations measured in urban park trees. Positive correlations between Cu, Pb and Fe concentrations in leaf tissues demonstrated the significance of deposition to leaf tissue content. This supports that aerial deposition to leaves is an important source of metal contamination in leaves. Further evidence was the lack of correlation between metal concentration in leaves and soil. Soil samples possessed a higher metal burden than leaves, which postulates that the soil was an appropriate indicator of long-term metal deposition but its utility is limited in the assessment of metals, which are highly mobile or major component of soil.

Concentrations of seven elements (As, Cd, Cr, Mn, Pb, V, Zn) in mosses (*Hylocomium splendens, Pleurozium schreberi, Eurhynchium angustirete*) and needles of Norway spruce (*Picea abies*) and juniper (*Juniperus communis*) were determined at 48 sites in Lithuania. Conifer needles consistently showed many times lower concentrations than mosses collected at the same site. Correlations between heavy-metal concentrations in needles and mosses indicated that accumulation processes may be similar, but mosses appear to be clearly preferable as biomonitors of atmospheric deposition because of their higher elemental concentrations and more quantitative reflection of deposition rates.

Effects of Environmental pollution from road traffic on trace element accumulation and deposition were examined in washed and unwashed *Petunia* leaves. Substantial amounts of elements were removed simply by washing with de-mineralized water, which removes at least 45% of Al, Fe, and Pb and 15% of Mn, Cu, Zn in urban areas, where the aerial deposition took place. Throughout the growth, the concentration of Fe, Cu, Al, Ni and Pb increased in the washed leaves of the petunia plants grown in urban areas. However, the plant controlled very specifically the contents of Mn and Zn. Concentrations of elements were significantly high in washed leaves from the urban areas than those from the suburban areas, indicating that this plant is able to absorb Fe, Ni, Al and Pb through its roots and leaves (Caselles *et. al.*, 2002).

Brief characteristic descriptions of commonly found heavy metals in the environment are given below:

Cadmium (Cd)

Cadmium is a nonessential metal to living organism and can become toxic by displacing Zinc (Leborans and Novillo, 1996). Cd content in surface soil is strongly influenced by man's activity. Naturally Cadmium occurs in all plants but has not been shown to be essential in plant nutrition or metabolism, and is probably taken up passively, and usually confined to roots (Streit and Stumm, 1993).

Plants vary in their sensitivity to Cd in nutrient solutions from 0.2 to 9 mg/kg. Cd at 3 mg/kg in plants will depress plant growth. Cd toxicity symptoms are leaf chlorosis and necrosis followed by leaf abscission as Cd interferes with net photosynthesis and the uptake and transport of mineral elements in plants. The effect of Cd on plants varies with species and with other elements, with both synergistic and antagonistic effects. Cd content in plants is correlated with that found in soils; the degree of Cd increase varies widely with plants species. Cd solubility in soil decreases significantly with an increase in soil pH. When exposed to high levels of Cd in rooting environment, some root crops and leafy vegetable will contain sufficient Cd to pose a potential health hazard in their consumption. For some seed crops, Cd is not translocated to thc developing seed., although the plant is exposed to high levels of Cd in rooting medium. The index of bioaccumulation ranges between 1 to 10. Therefore, Cd will accumulate in some portions of the environment, posing a potential health hazard due to its uptake and accumulation in some food crops. Cadmium accumulation in plants varies with the pH of the soil (Peralta-Videa *et. al.*, 2002).

Cd causes chlorosis, browning of leaf margins, reddish veins and petiole, curled leaves and brown stunted roots. Reduced conductivity of stem is caused by the deterioration of Xylem tissues and reduced growth of roots and number of tillers in cultivars (Kabata-Peadias and Pendias, 1994).

Crop growth decreases with increasing rates of cadmium application. Phytotoxic symptoms on wheat leaves were observed only at 50 and 100ppm Cd treatments. These assume the form of light whitish yellow spots in the interveinal areas, later these spots coalesce and turn yellowish to brownish in colour, leaf tips also show marginal scorching (Singh *et. al.*, 1991). Gupta and Dixit (1992) found that crop's tolerance to Cd toxicity depends on the crop's ability to absorb Cd.

High Cd content decreased grain formation in pea (Sehgal and Ratan, 1990). Gupta and Potolia (1990) found that application of Zn enhanced wheat grain and straw yield, while cadmium application caused a drastic reduction in yield. The concentration of Zn and Cd in grain and straw increased with their respective application. The Zn concentration in the seeds was enhanced as well. Addition of compost, lime and phosphorus reduces the toxicity of Cd in rice (Sarkuman *et. al.*, 1991).

Sarkuman *et. al.* (1991), conducted a field experiment to study the toxicity of Cd to the rice grown under flooded conditions in an alluvial soil. The study suggested that in water logged soils rich in ferrous iron, Cd uptake by rice and Cd phyto-toxicity are unlikely. The process of transformation in such soil appears to be mainly governed by the formation of Fe^{+2}, which favors the reversion of soluble Cd to form inert mixed oxides of Cd and Fe. It is also evident that the reoxidation of Cd does not occur during the normal cultivation practice followed for low land rice.

Xian (1990) with his experiments on the distribution of Cd in the root tissue of the Cd tolerant plant *Athyrium yokoscence*, suggested that the Cd in cells is strongly associated and settled in the cell wall. This association weakens or eliminates Cd's harmful effect on the metabolic system and thus maintains normal plant growth in high soil levels of Cadmium.

Norhao and Gaur (1995) reported that cations including calcium, magnesium, potassium, sodium, copper, iron, nickel and zinc inhibited the extracellular binding and intracellular uptake of Cd by *Lemma polyrhiza* in solution culture. They also reported that Cu, Ni and Zn competitively inhibited the extracellualr binding of Cd. Fe and Na caused mixed inhibition of extracellular binding. Ca, Mg, Fe, Ni and Zn competitively inhibited intracellular Cd uptake. These studies also suggest that a high level of cations and metals in external environment should be expected to lower the Cd accumulation efficiency of Lpolyrhiza.

Chromium (Cr)

Cr content in plants ranges from 0.02-0.2 mg/kg with phytotoxicity at >10 mg/kg. Cr content in air ranges between 0.001-1.0 mg/m3, in industrial areas values may reach upto 30-50 mg/m^3. Cr has a moderate index of bioaccumulation (Pais and Jones, 1997).

In plants Chromium combines with large molecules mainly proteins, including enzymes with catalytic properties (Kabata-Peadias and Pendias, 1994).

According to Merkert, (1994) normal Cr content in plants is 1.5 mg/kg but 0.1-0.5 mg/kg (dry weight) of Cr in leaf tissue is toxic for the plant (Kabata-Peadias and Pendias, 1994). Chlorosis of new leaves, necrotic spots and injured root growth was caused by Chromium (Kabata-Peadias and Pendias, 1994).

Plants vary in their sensitivity to Cr. Chromium tends to accumulate in roots and is not easily translocated. There is some evidence of stimulatory effect of low levels of Cr on plants growth.

Protein, pigment and enzyme analysis were conducted in both Cr-tolerant and Cr-sensitive cultivators of mungbean after 72h of treatment (Samantary, 2002).. Chlorophyll and protein contents were reduced in Cr-sensitive cultivators more than those of the tolerant ones. The enzyme activity varied among the Cr-tolerant and Cr-sensitive ones. Activity of peroxidase was greater in Cr-sensitive than tolerant cultivators. Tripathi and Sadhna (1999) also observed a decrease in chlorophyll and protein content in *Albizia lebbek* leaves on Cr treatment.

The Cr^{+4} form is toxic to plants, animals and man, although its occurrence is not common in the environment. Cr^{+4} is 100 times more toxic to plants, animals and man than trivalent chromium. The solubility of both forms is significantly affected by pH, the lowest solubility occurs between pH 5.5 and 8.0. Plant availability of Cr on high content soils can be reduced by liming and by the addition of phosphate fertilizer and organic matter.

Copper (Cu)

Cu has a high bioaccumulation index and can accumulate in plants to high levels, which may pose environmental health problems (Pais and Jones, 1997). Cu exists in soil primarily in complex forms as low molecular weight organic compounds such as humic and fulvic acids. Cupric ion is present in very small quantities in soil solution with Cu deficiency occurring primarily on sandy and organic soils. Cu uptake rates are lower than for other elements.

Soluble content in soil ranges from 3-135 mg/L. In plants Cu content varies from 1 to 10 mg/kg. Cu is immobile in soil. It is relatively uniform in profile distribution, although there can be considerable bioaccumulation at the soil surface. Cu can be easily precipitated and interacts readily with both organic and inorganic substances with widely varying solubility's due to the pH. In cropland soil Cu contamination is not uncommon in areas where Cu containing chemical's were in common use as well as from industrial pollution and or from land deposition of industrial wastes and sewage sludge. Decrease in plant availability can be obtained by liming, the addition of phosphate fertilizer and addition of organic matter to the soil.

Copper is the constituent of chloroplast protein plastocyanin, as well as serving as a part of electron transport system linking photosystem I and II, This element participates in protein and carbohydrate metabolism and N_2 fixation. It is a part of the enzymes cytochrome oxidase, ascorbic acid oxidase which reduce both atoms of molecular oxygen. Copper is also involved in the desaturation and hydroxylation of Fatty acids. Copper sufficiency range in leaves is between 3 and 7 mg/kg of the dry matter, while the toxicity range begins at 20-30 mg/kg. Much higher values, 20 to 200 mg/kg can be tolerated if cooper has been applied as a fungicide (Pais and Jones, 1997).

Copper is also the main constituent of various oxidases and plastocarbohydrate in plants. Cooper is mainly associated with oxidation, photosynthesis, protein and cyanins metabolism (Kabata-Peadias and Pendias, 1994). Cu causes changes in the permeability of cell membrane (Pais and Jones, 1997). Cu in leaf tissue is toxic for the plant (Kabata-Peadias and Pendias, 1994).

Cu is an essential micronutrient with a sufficiency range of 5-30 mg/kg dry weight of tissue. Plant species vary in their tolerance to Cu. Toxicity occurs when Cu tissue levels exceed 20-30 mg/kg. Cu is particularly confined to roots, with relatively little being translocated to plant tops. The symptom of Cu toxicity is Chlorosis, as high Cu interferes with Fe metabolism. Cu deficiency is not common because the requirement for most crops is quite low. Cu deficiencies are most likely to occur on organic soils with high pH (>7.5) and or high organic matter content.

Dark green leaves followed by Fe induced Chlorosis, thick short or barbed wire roots and depressed tillering was caused by Copper in cultivars (Kabata-Peadias and Pendias, 1994). Copper in plants can interfere with iron metabolism, which may result in the development of iron deficiency. In its interaction with the molybdenum it may interfere with the enzymatic reduction of nitrate. An excess of Cu can induce iron deficiency and chlorosis. Root growth may be suppressed with inhibited elongation and lateral root formation.

Iron (Fe)

Iron is a major constituent of lithosphere. Its geochemistry is very complex, depending on its valence state. Iron is associated with various physicochemical conditions. In soil Fe exist mainly as oxides and hydroxides and is chelated by organic matter. Iron in soil solution is affected by soil pH, decreasing with increasing pH. Under anaerobic conditions, ferric (Fe^{+3}) is reduced to ferrous (Fe^{+2}), which significantly increases its solubility in soils and under acid soil conditions induces possible iron toxicity. For humans and plants, the essentiality of iron was demonstrated over two centuries ago. Abundance of Iron in lithosphere is 41,000 mg/kg with common valence states are Fe^{+2} and Fe^{+3}. Total iron content in soil is 38 mg/kg. Fe content in plants varies from 20-100 mg/kg (Pais and Jones, 1997, Dara, 1998).Average concentration of Iron in air is 0.001 mol/m^3 and examples of functions in cell are energy transfer proteins, co-enzyme factor prosthetic group (Porter and Lawlor, 1991). Iron has a low bioaccumulation index. Iron is mainly involved in photosynthesis, N_2 fixation and an important constituent of proteins, dehydrogenases, and Ferrodoxins (Kabata-Peadias and Pendias, 1994).

Iron is the important component of many plant enzyme systems, such as cytochrome oxidase and cytochrome. Iron is a component of protein, ferredoxin and is required for nitrate and sulphate reduction, nitrogen assimilation and energy production. It functions as a catalyst or part of an enzyme system associated with chlorophyll formation. It is thought that iron is involved in protein synthesis and root tip meristem growth.

Leaf iron content ranges between 10 to 1000 mg/kg in dry matter, with sufficiency ranges from 50 to 75 mg/kg, although total iron may not always be related to sufficiency. In general, 50 mg/kg iron level is the generally accepted critical value for most crops. In majority of plants iron is in the Ferric form, as ferric phospho protein, although the ferrous ion is believed to be the metabolically active form. High Zn can interfere with iron metabolism, resulting in visual symptom of iron deficiency (Pais and Jones, 1997).

Iron may accumulate to several hundred milligrams per kilogram without symptoms of toxicity. Toxicity produces a bronzing of leaves with tiny brown spots.

Fe is an essential micronutrient. Its sufficiency range in most plants is 50-500 mg/kg dry weights; the critical level is 50 mg/kg for a wide range of plants. Total iron, is not however an indicator of sufficiency and extractable iron is frequently used for iron status assessment, where 20-25 mg/kg is the critical range, depending on the extraction procedure used and plant species. An accurate assessment of the iron status of plant can be significantly affected by the presence of extraneous iron on tissue surfaces due to aerial deposition and soil contamination as well as contact with iron containing devices used for tissue preparation. In many instances indirect measurements such as chlorophyll content determination, may be more consistent evaluator of iron status of plant. Fe deficiency affects many crops and may be the most frequently occurring deficiency worldwide. Deficiency is mainly associated with alkaline soil condition, referred to as lime Chlorosis. Iron deficiency may also be genetically controlled, as some plants have been classed as either iron efficient or iron inefficient. So called iron efficient plants are able to acidify the rhizosphere and release iron chelating substances called siderophores. Zn and presence of bicarbonate anion in the soil solution can induce iron deficiency. Once it occurs, iron deficiency in plant is difficult to correct even with frequent foliar application of iron containing compounds, whether chelated or not. The biochemistry of iron is very complex. Fe is absorbed by leaf cuticle and roots (Caselles *et. al*., 2002).

The ferric form of iron in soil, whose availability is affected by the degree of soil aeration, is thought to be the active form taken up by plants. Iron sufficient plants can acidify the rhizosphere as well as release iron complexing substances, such as siderophores, which enhance availability and uptake. Iron deficient plants do not have roots that will readily acidify the rhizosphere and or release iron complexing substances.

Manganese (Mn)

Major sources of Manganese in urban environment are gasoline and municipal solid waste.In plants Mn is mainly involved in photoproduction of oxygen in chloroplasts and indirectly on NO_3 reduction (Kabata-Peadias and Pendias, 1994). Normal Mn content in plants is 200 mg/kg and 20-300 mg/kg (d.w.) of Mn in leaf tissue is toxic for the plant (Kabata-Peadias and Pendias, 1994).

Manganese is one of the most abundant trace elements in the lithosphere. It exists in the soil in a wide range of oxidation states, depending on the physicochemical conditions, and its solubility is dependent on soil pH. Mn is not evenly distributed in the soil profile. An excess of Mn causes older leaves to show a brown spot surrounded by a chlorotic zone

Manganese is an essential element for plants. Manganese toxicity is frequently associated with low (<5.5) soil pH. Manganese availability can be reduced significantly by low soil temperature. Mn is not known to interfere with the metabolism or uptake of any other element.

Plants vary considerably in their sensitivity to Mn and plants sensitive to deficiency are equally sensitive to toxicity. Some plants are high accumulators of Mn without detrimental effects. In general as the soil pH increases, the availability of Mn decreases, as manganese deficiency is related to high soil pH (>7.5) and toxicity is associated with low soil pH (<5.5). Low soil temperature and increasing soil organic matter content also reduce Mn availability. Mn deficiency symptoms occur primarily on emerging leaves and can be frequently confused with iron deficiency. Mn toxicity can occur on acid soils (pH<5.5) and under anaerobic conditions. It accumulates as specks of manganese oxide in the epidermis, giving a symptom referred to as measles. The manganese status of plant can be significantly influenced by its relationship to other elements, primarily iron, as well as phosphorus, calcium, magnesium and silicon.

Progressive increase in $MnSO_4$ concentration up to $5x10^{-3}$ M in Mungbean brought about a progressive decrease in total chlorophyll and chl a and carotenoids content. Chl b content changed very little by excess manganese treatment Sinha *et. al.* (2002).

Zinc (Zn)

Zinc is released in urban environment (air and soil) through combustion of fuel, burning and dumping of municipal solid waste and other commercial activities.Zinc is an essential micronutrient. The critical zinc value has been accepted as 15 mg/kg for most crops, although under some conditions 10 mg/kg may be sufficient. Zn tends to accumulate in older leaves. A typical symptom of Zn deficiency is resetting of terminals. Zinc has a fairly high bioaccumulation index.

Zn deficiency is most likely to occur on leached sandy soils low in organic matter content, neutral to alkaline soils, and or soils high in available phosphorus. The effect of phosphorus on Zinc is thought to occur within the plant and is not a major factor in zinc uptake. Other important interactions with copper, iron, arsenic and nitrogen affect zinc metabolism. A number of major food crops such as corn, sorghum, citrus, and field bean are particularly sensitive to zinc. Toxicity can occur when soil zinc levels are elevated by the application of industrial waste or sewage sludge, which can contain zinc levels in excess of 17000 mg/kg. The effects of zinc toxicity can be significantly reduced by liming the soil to pH 6.0 or above.

In plants Zn is mainly involved in carbohydrate and protein metabolism (Kabata-Peadias and Pendias, 1994). Zinc availability is affected by soil pH, decreasing with increasing pH. Zn availability can also be reduced when the available soil phosphorus level is very high. Zn is brought into contact with plant roots by mass flow and diffusion, with diffusion being the primary delivery method. Cu^{+2} and other cations such as the ammonium cation will inhibit root Zn uptake. The efficiency of Zn uptake seems to be enhanced by a reduction of the pH in the rhizosphere. Those plants species that effectively reduce the pH are less affected by high soil zinc than those plant species that do not.

High soil Zn content significantly reduces seed germination in some crops. There is evidence that Zn sometimes depresses Cu uptake from soils, this is thought to be due to a competitive interaction during uptake rather than to a soil effect (Katyal *et. al.*, 1992).

Addition of phosphatic fertilizers has caused Zn deficiency in many crops. It was observed that Zn concentrations in wheat and Soyabean decreased with increasing levels of phosphorus. (Rathore *et. al.*, 1995).

High Zinc content in plant leaves caused chlorosis in new leaves, necrosis in leaf tips, interveinal chlorosis in new leaves, retarded growth of entire plant and injured roots (Kabata-Peadias and Pendias, 1994).

Zn is involved in same enzymatic reactions as are by Mn. Some plants can accumulate sufficient quantities of Zn (several hundred mg/kg) without any injury to the plant. High Zn content in soil can induce iron deficiency in plants. High phosphorus in soil can interfere with Zn metabolism as well as with Zn uptake by the roots. Plants particularly sensitive to iron will become chlorotic when Zn levels are abnormally high.

Lead (Pb)

Presence of Pb in the environment has led to increasing awareness and concern due to its detrimental effect to ecosystem and human health (Mathur *et. al.*, 1997). Abundance of average lead concentration in lithosphere is 14 mg/kg and common valence state Pb^{+2}. Total content in soil ranges from 3-189 mg/kg with natural background level 10-67 mg/kg (mean 32 mg/kg), and 100-400 mg/kg in soil considered phytotoxic. Total soluble content in saturated paste of soil is 5.0 μg/L. Lead content in a reference plant is 1.0 mg/kg. Lead has a moderate bioaccumulation index and its movement within the food chain is due primarily to man's activity (Pais and Jones, 1997).

Lead is not required by any living organism and is highly toxic in nature, commonly found in urban areas because it is added to gasoline, a prime fuel in urban transport sector, to raise octane levels and help engine run more smoothly. It is emitted as tiny particles in exhaust contaminating air, soil and food. Lead (Pb) and some of its chemical compounds are virtually ubiquitous in the environment. Lead can be found in air, drinking water, as well as in soil and various food items. An estimated 80-90% of lead in ambient air is derived from the combustion of leaded petrol (Smith, 1981 and Jain *et. al.*, 1986). The degree of pollution from this source differs from country to country, depending upon motor vehicle density and efficiency of efforts to reduce the lead content of petrol (WHO 1987e). About 1% of lead in the petrol is emitted unchanged as tetralkyl lead (organic lead). Concentration of tetralkyl lead amounting to more than 10% of the total lead in the ambient air has been measured in the vicinity of the service stations (NSIEM, 1983). The WHO guidelines values for long term exposure (*e.g.* annual average) to lead in the air is (0.5 to 1.0$\mu g/m^3$ (WHO, 1987e). Majority of lead in ambient air is as fine particles (μm). The possibility of lead exposure in human is therefore of great significance from health point of view. Lead does not spare any organ in the body and is not known to serve any necessary biological function.

Lead is not bio-degradable, never disappears, only accumulates where it is deposited, provides no known biological benefits to humans, young children absorb lead more rapidly than adults, more than 95% of the retained lead is accumulated in bones, where it is in continuous exchange with soft tissue pools, the half - life of circulating lead in blood is about one month (Dara, 1998).

Lead is least mobile among other heavy metals in soil. It accumulates primarily on the surface soil, where its increasing presence may begin to affect soil micro flora. Lead availability in soil is influenced by pH, decreasing with increasing pH of the soil. Lead is not readily soluble in the water and is found in relatively low concentrations (Pais and Jones, 1997).

Urban soils were enriched with anthropogenic lead such as industrial and automobile lead because anthropogenic lead was relatively stable after deposition in soils (Wong *et. al.*, 2002). Due to high affinity for soils lead is estimated to have long retention times in some cases for as long as 5000 years (Friedland, 1990).

A toxic heavy metal now present in the petrol called "super", which is needed to ensure problem-free working for old engines. At the present time the maximum level of lead present in that fuel is set at 0.15 grams of lead per liter of petrol. "lead" - in the air have decreased dramatically since 1978, primarily due to reductions in emissions from automobiles. Today, metal processing plants are generally responsible for most of the lead in the air (Published by the US EPA Office of Air Quality Planning & Standards November 2000).

Lead is added to petrol as tetra-alkyl lead and ethyl-trimethyl lead (Singh *et. al.*, 1997). Lead content of petrol differs in different countries *viz*. Bangkok 0.15 g/l (1992), Beijing 0.4 - 0.8 g/l, Cario 0.8 g/l, Delhi 1.8 g/l, Kolkata 0.1 g/l, London 0.15 g/l, Los Angeles 0.026 g/l, Mumbai 0.15 g/l (1991), New York 0.026 g/l, and Tokyo 0.15 g/l (WHO/UNEP, 1992).

In **Los Angeles**, petrol is primarily Pb free and little leaded petrol sold contains only 0.026 g/l Pb.

In **Moscow**, prohibition on the sale of leaded petrol and reduction in motor vehicle traffic has helped to reduce Pb levels since the late 1980s (Rovinsky *et. al.*, 1991). The Pb content in air during the period 1985-1990 varied from 0.01-0.04 $\mu g/m^3$. The levels are well below the WHO guidelines (1 $\mu g/m^3$). In Japan, the maximum level of lead additive in petrol is 0.15 g/l. Pb emissions were reduced when leaded petrol was introduced.

In **Bangkok** lead in commercial petrol was reduced from 0.84 g/l to 0.45 g/l in 1984 and to 0.40 g/l in 1989.

In **Beijing**, there is a0.8 g/l limit for lead content in petrol, and 80% of the petrol consumed is lead free. Furthermore traffic in Beijing is still limited, so emissions of Pb are small and well below 100 tonnes per annum (Vahter and Solarch, 1990).

In **Cairo**, thc maximum Pb content of Petrol is 0.8 g/l which is relatively high. It may be estimated that a motor vehicle population of 900,000 could produce Pb emissions of the order of 1000-2000 tonnes per annum. Pb levels exceed the WHO annual mean guideline all over the Cairo (Ali *et.al.*, 1986). Pb concentrations were greatest in the summer *e.g.* 64 $\mu g/m^3$ in June 1984.

Lead concentrations in petrol in Pakistan are between 1.5 and 2.0 g/l, which are relatively high on the world scale. Concentrations of the inorganic lead are very high in Karachi. In Rio De Janerio (FEEMA, 1989) and Sao Paulo (CETSEB, 1989), the Pb levels are low because of reliance on alcohol-powered vehicles coupled with reduced Pb levels in petrol.

In **India**, Pb addition amounts to 150-300 mg Pb/l, of which 80-90% escapes in car exhaust. Motorcars release approximately 80 mg Pb/ km traveled (Smith 1981). The concentration of lead found in various environmental compartments bordering a roadway is a function of a number of factors including distance and meteorology (Mielke 1991) and 25% of vehicle emitted Pb is coarse grained and thus deposited close to the road. The remaining 75% is fine and may remain air borne (Fergusson and Kim, 1991)

In **Mumbai**, petrol driven vehicles are the main source of lead in air. Vehicle number has risen from 125000 in 1971 to 468000 in 1987 and to 588000 in 1989. In 1986 the Pb content of petrol in Mumbai was 0.8 g/l for premium and 0.56 g/l for regular. The mean Pb content of petrol from Mumbai refineries in 1991 was 0.155 gl^{-1} (NEERI 1991b). Monitoring indicates that annual air borne Pb levels have fallen significantly since 1970s to between 0.02 and 0.33 $\mu g/m^3$, well below WHO guideline of 1$\mu g/m^3$.

Lead content of petrol from the Haldia refinery, which supplies petrol to Kolkata, is 0.1 g/l. Despite the relatively low lead content of petrol, the annual air borne lead levels were highest in India (NEERI, 1991d). The annual concentrations are highest in the residential and commercial sites (0.73 $\mu g/m^3$), but below 1$\mu g/m^3$.

The Pb content of petrol distributed from Mathura refinery which serves Delhi is relatively high (1.8 g/l). Given this Pb content and the number of petrol driven motor vehicles registered in Delhi, it is estimated that Pb emissions from this source are of the order of 600 tonnes per annum (NEERI, 1991a).

Recognizing the health hazards of lead almost all industrial countries have progressively reduced lead content of petrol over the past decade.

In developing countries reduction of lead level in petrol has not changed ambient lead concentration because of continuous increase in vehicle number. With more vehicles on the roads, lead reduction in petrol has become a major priority in developing countries. Severe lead pollution from single point sources can expose large number of people living in the area to excessive lead levels.

Akhter and Madany (1993) have estimated very high *i.e.* 697.2 mg/g Pb in street dust and 360 mg/g in the house dust of Bahrain, which was highest among all the toxic metals, detected. He also summarized level of the metals in the dust of various countries like Greece, Kenya, Kuwait, Nigeria, Scotland, and Taiwan have lesser Pb^{+2} in the environmental dust ranging from 4.1-9.50 mg/g dust whereas countries like UK, USA, Canada, Germany, Malaysia, Netherlands and New Zealand have a maximum level upto 17.9 mg/g in certain areas.

Study by Gajhate and Hassan (1999) reveals that Pb concentration in respirable dust in major cities of India was Ahmedabad 0.009-0.26, Mumbai 0.048-0.31, Kochi 0.016-0.21, Kolkata 0.43-2.30, Delhi 0.05-0.45, Hyderabad 0.01-0.28, Kanpur 0.5-2.23, Chennai 0.02-0.21and Nagpur 0.02-0.16 ìg/m^3. Pb concentration was highest in winter season followed by summer and monsoon season in most of the Indian cities.

Lead at 30 mg/L in nutrient solution has been found to be toxic to plants with 10mg/L slowing plant growth and 100mg/L being lethal. In some types of plants, lead can be as high as 350 mg/kg in plant tissue without visible harm. Lead can be readily absorbed by plant roots (the amount absorbed varies greatly with plant type), but little (less than 3%) is translocated to the tops. Background lead levels in grasses are 2.1 mg/kg and 2.5 mg/kg in clovers. Leafy vegetables, such as lettuce have a high accumulation character for lead.

Pb causes changes in the permeability of cell membrane (Pais and Jones, 1997). According to Merkert, (1994) normal Pb content in plant was 1.0 mg/kg and 5-10 mg/kg (dry weight) of Pb in leaf tissue is toxic for the plant (Kabata-Peadias and Pendias, 1994). Lead causes wilting of older leaves, stunted foliage, darker color of green leaves and shortening of roots (Kabata-Peadias and Pendias, 1994).

Plant content of lead contamination depends upon several factors, including distance from road, nature of collecting surface of plant, duration of exposure, traffic density and direction of wind. Pb contamination follows motorway areas. Roadside vegetation may have levels of 50ppm Pb, but at a distance of only150m from the highway, the level is about 2 or 3ppm (Kafka and Kuras, 1997).

Translocation of lead to the upper ground portion of a plant is partially defined by the plant root system. Stefanov *et. al.* (1992) reported that 90% of the total lead accumulates in the roots of *Phaseolus vulgaris*, whereas in *Zea mays* (Stefanov *et. al.*, 1993) only 45% of the total lead is incorporated in the roots and the remaining 55% gathers in the leaves. In the upper ground parts of *Capsicum annum*, lead accumulation mainly occurs in the leaves (Stefanov *et. al.*, 1995).

Plants absorb lead from the soil as well as from the air (Singh *et. al.* 1997). Susceptible plant species growing in lead enriched environment are able to accommodate large quantities of lead in their organs (Kumar *et. al.* 1993, Bharti and Singh 1993). Plants of *Cassia tora* and *C occidentalis* growing by the roadside accumulate upto 300 mg/g dry wt. of lead (Krishnayapa and Bedi, 1986).

Photosynthesis is considered to be the most sensitive metabolic process of lead toxicity. The metal inhibits the rate of photosynthesis in different plants (Singh *et. al.* 1997). At 193 μM leaf lead, the rate of photosynthesis was reduced by 50%. This reduction may be due to the closing of stomata, disruption of the chloroplastic organization, change in metabolites of photosynthesis, replacement of essential ions like Mg^{+2}, Mn^{+2} etc by Pb^{+2} in the chloroplast, inhibiting de novo synthesis of photosynthetic pigments *i.e.* chlorophyll and carotenoids as well as acceleration during in vivo degradation of pigment molecules (Kumar *et. al.* 1993).

Naturally Lead occurs in all plants but has not been shown to be essential in plant nutrition or metabolism, and probably taken up passively, and usually confines to roots (Streit and Stumm, 1993).

Polyalthia longifolia leaves showed maximum accumulation of lead in high-density automobile traffic area, hence it is sensitive plant and respond to atmospheric lead levels (Naik and Deshpande, 2000).

The total soluble protein in the leaves of *Sesamum indicum* increases from 4.79 to 8.26 μg protein/gm fresh weight as the lead supply was increased from 0.00 to 0.5μM (Kumar *et. al.*, 1993). The maximum lead was invariably found in the foliage of plants growing along the road with high traffic density and the minimum amount of lead was collected in plants along roads with minimum traffic load (Singh *et. al.* 1995).

A study was carried out by Singh *et. al.,* (1995) along ten road transactions of Lucknow city for estimation of Pb. The road transaction around Alambagh (high traffic density) was highly polluted with lead (2.96 $\mu g/m^3$) in suspended particulate matter (1080 $\mu g/m^3$). With increase in traffic density, there was corresponding increase in ambient lead contents being minimum (0.46 $\mu g/m^3$) in Vikramaditya Marg (lowest traffic density) and maximum (2.96 $\mu g/m^3$) in Alambagh (highest traffic density).

The changing levels of lead in the soil and vegetation along two national highways near Lucknow, India were investigated by Singh *et. al.* (1997). The pattern of Pb deposition, as reflected by soil burdens, shows decrease in concentration with increasing distances from road margins, which reflect a decrease in rate of aerial deposition. At both sites Pb concentration was above background concentration even at the soil core depth of 15 cm. Pb deposition at the road edge was 44 μg/g, which decrease threefold between 60 m and 80 m from the roadside edge. Lead although not readily soluble in soil, is absorbed mainly by root hairs and is stored in cell walls. The translocation of Pb from root to tops is greatly limited (Nwosu *et. al.*, 1995). The concentration of Pb in plant leaves was high in comparison to other parts of plant and concentration showed a decreasing trend as the distance increased from the road edge (Singh *et. al.*, 1997).

Chapter - 9

MITIGATORY METHODS TO URBAN AIR POLLUTION

Through the sharing of experience of developed countries, which faced the massive health problem due to vehicular pollution in the past and remedial measures adopted by them, it is needed to avoid mistakes made in the past and instead introduce effective measures for the future to reduce or confine the damage that has already been incurred. A progressive policy of motor vehicle emission control strategy, which may be more feasible/affordable, has to be adopted for solving immediate air pollution problems. Uncontrolled growth of vehicular fleet and its emission is taking place in the urban areas, which must be controlled/regulated in order to prevent future disaster. Because of economic development and speed of motorization expected over next decades this problem will become more vulnerable and even greater, If control measure are not adopted timely. It is suggested that planning must begin to provide alternatives to the motor vehicles and to reduce emission of the vehicles. The country without high capital resources will face challenging problems for evolving control strategy that will be acceptable both economically and socially.

Mitigation Strategy

The approaches should be made in a way that should not affect our developing economy. However, simultaneously a policy should be strictly followed in order to protect the environment from automobile emissions. Progressive strategy includes first alternative arrangement then the policy should be forcibly imposed to save the deterioration. The economist and environmentalist should find out mutually beneficial solutions to over come and develop the economic status. The goal of sustainable mitigation strategy will be balancing both the economy as well as the environment. These approaches are generally termed as green economics, green technology and green planning *etc*. Goal can be achieved in the following ways:

1. Natural measures
2. Vehicle measures
3. Transport management and policy
4. Moral and education awareness

(1) Natural Measures

The deleterious effects of air pollutants on plants have long been recognized. Several studies here revealed that certain plants can absorb the gaseous pollutants from atmosphere directly. The rate of removal of gaseous pollutant were in the following order: Hydrogen fluoride (HF), Sulphur dioxide(SO_2), Nitrogen dioxide (NO_2), Ozone (O_3), Peroxyacetyl nitrate (PAN), Nitric oxide (NO) and Carbon monoxide (CO) (Hill, 1971). Vegetation could therefore be an important sink for the pollutants (Garland *et al*, 1973; Siebke *et al*, 1990). Since plants act as scavenger of several pollutants, this property of plants can be judiciously used in reducing emission from the environment. Recent studies pertaining to the role of plants in monitoring and mitigating air pollution have shown that certain plants can be used to reduce the emissions. However, it seems to be the long way to achieve goal but indeed is a novel approach, besides being economically acceptable and affordable control technology. The following plants have been studied specifically and were found to possess extraordinary capability of reducing auto emission. Ailanthus excelsa Roxb., *Antigonor leptopus* Hook., *Azadirachta indica* A. juss., Bougainvillea sp., *Callistemon lanceolatus* R. Pr., *Calotropis procera* R. Br., *Cassia fistula* L., *Clerodendron splendense* Linn., *Dalbergia sisso* Roxb., *Delonix regia* Rafin., *Ficus religiosa* L., *Holoptelea Integrifolia*., Planch., *Lantana camara* L., *Moringa oleifera* Lam., *Pithecolobium dulce* Benth., *Polyalthia longifolia* Benth. & Hook., *Pongamia glabra* vent., *Ricinus communis* L., *Tabernaemontana coronaria* willed. and *Thevetia nerifolia* juss.

Green belt should be designed and developed in and around cities. The scavenging properties of plant must be exploited in making green cities. This way will be economically and environmentally acceptable for urban centres in order to be a sustainable city in future.

2. Vehicle Measures

(A)Technological Ways

I). Emission factor for vehicle registration:

Registration of vehicle is based on the emission standards. Only those vehicles should be allowed to register which fulfill the emission standard criteria. The old and poorly maintained vehicles should be phased out from city gradually by giving definite time frame.

II). Clean technology

(Catalytic converter, unleaded petrol, evaporative emission control system, exhaust gas recirculating system and ignition timing *etc.*) based vehicles', high occupancy vehicles (HOV) should only be allowed on the road in the future. Catalytic converter and other filters should be made compulsory for new vehicles as well as old. Some of the clean technologies are:

a) Lead free petrol

In petrol driven engine, the mixture of air and fuel is introduced by aspiration, pressurise and lit by a high voltage electrical charge. So as to prevent this mixture from exploding in the compression stage in the cylinder, a substance must be added to prevent detonation. Traditionally the most widely used one has been tetraethyl lead. The problem is that combustion produces lead and lead oxide, which are emitted in the atmosphere through the exhaust pipe and have a pollutant effect as well as possible effects on health. The use of lead free petrol prevents the emission of lead and allows the use of catalysers as well as reducing the level of pollution given out.

b) Three way catalysers

A mixture of gases is obtained on the combustion of fuel. Pollutant gases are liable to produce photochemical smog in the presence of environmental humidity and highly corrosive nitrogen acid. To prevent this, three way catalysers are used. The three-way catalyser is a container that is placed inside the exhaust pipe and which consists of ceramic sponge fibre lined with certain noble metals such as silver, rhodium or palladium. These metals transform the highly reactive pollutant gases into the gases already present in the natural makeup of the atmosphere. So the nitrogen oxide, carbon monoxide and the hydrocarbons are turned into nitrogen, carbon dioxide and water.

In order to ensure the proper working of the catalyser, the combustion inside the engine must be carried out without too much or too little oxygen. Improper use of catalyser may increase emission of pollutants.

III). Vehicle maintenance

Vehicles should be maintained according to the manufacturer's recommendations (*e.g.* Tune ups, replacement of spark plug, carburetor adjustment *etc*).

IV) Inspection, maintenance and antitempering checks:

An effective programme of inspection, maintenance and antitampering should be regularized so that the vehicles are replaced/ repaired if tampered for better mileage. The owner should be benefited by incentives or charged penalties according to vehicle's condition.

(B) Fuel ways

Energy supply options that increase the efficiency of producing energy carriers primary emissions of green house gases can contribute to sustainable development objectives.

i) Compressed natural gas

Natural gas has lowest emission rate of all the fossil fuels in general, and should be used more efficiently than the fuel supplied.

ii) Electric driven vehicle for transport

Electric driven vehicles have great potential for dramatic reduction of air pollutants emissions and marked improvement in fuel economy.

iii) Clean fuel for transportation

Alternative transportation fuels are important now *e.g.* hydrogen offers good prospects for simultaneously dealing with the multiple challenges facing energy system in 21st century. Hydrogen is a clean versatile and easy to use energy carrier. It can also be derived from fossil fuel. Hydrogen used in fuel cell vehicles would emit significantly lower CO than gasoline internal vehicle combustion.

Bio-fuels

Bio-fuels are obtained from agricultural products and have a heat power similar to that of fossil fuels, which allows them to be used in engines without making any large alterations to them. Bio-fuels do not contain sulphur and neither do they increase the amount of CO_2 given out into the atmosphere. The main problem facing bio fuels at the present time is their high production costs and their competitiveness compared to fossil fuel,

iv) Mix fuel technology

Mix fuel technology should be adapted in the automobile system to achieve the same or similar transport services.

v) Improving automobile fuel

The quality of fuel used in India is the worst in the world. It is therefore, essential to upgrade the quality of fuel through a reduction of sulphur, tetraethyl lead and aromatics and an increase in the octane number. The unleaded petrol should be encouraged. Hydrosulphurisation units for reduction, of the sulphur contents and catalytic reformer units for lead free petrol are required to be installed in the refineries. An increase in the octane number by 3 to 5 units will result in fuel economy and improvement in vehicular exhaust quality. This is possible through reduction of aromatics in the refining process Lowering aromatics will also help in reducing emission of volatile organic compounds (VOCs) during combustion. The CPCB had recommended the following specification for petrol.

Specification recommended by CPCB for diesel and petrol

Type of fuel	Content	Concentration
Diesel	Sulphur	0.5%
	Octane number	45
Petrol	TEL	0.15 gm/l
	Sulphur	0.15%
	Aromatics	25% by volume (maximum)
	Benzene	11% by volume (maximum)

vi) Programmes by government

Gasoline Benzene Reduction Program In India

Period	Benzene Content	Area Covered
Before 1996	No specification	Entire country
April, 1996	5%	Entire Country
April, 2000	3%	Metropolitan Cities
November, 2000	1%	National Capital Territory (NCT) and Mumbai

Source: Air quality status and trends, CPCB Report NAAQMS/14/2000-2001

Gasoline lead phase out programme in India

Phase	Effective from	Standard	Effective in
Phase I	June 1994	Lowleaded (0.15 g/l)	Delhi, Mumbai, Kolkata, Chennai
Phase II	1.4.1995	Unleaded (0.13g/l) + low leaded	Entire country
Phase III	1.1.1997	Lowleaded (0.15g/l)	Entire country
Phase IV	1.9.1998	Ban on leaded fuel	National Capital territory, Delhi
Phase V	31.12.1998 (advanced to 1.9.1998	Unleaded (0.13g/l) + low leaded	All others capital of states/union territories and other major cities
Phase VI	1.9.1999	Unleaded only (0.13g/l)	National capital region
Phase VII	1.4.2000	Unleaded (0.13g/l) + low leaded	Entire country

Diesel sulphur phase out programme in India

Phase	Effective from	Standard	Effective in
Phase I	April 1996	Low sulphur (0.5%)	Delhi, Mumbai, Kolkata, Chennai & Taj trapezium
Phase II	August 1997	Low sulphur (0.25%)	Delhi and Taj trapezium
Phase III	April 1998	Low sulphur (0.25%)	Metro cities
Phase IV	April 1999	Low sulphur (0.25%)	Entire country

(3) Transport Management And Policy

Effective traffic management and implementing clean air policies can also check emission.

i) Smooth flow of traffic

Smooth flow of traffic is only possible when road quality is improved. The potholes, speed breakers, encroachments and narrow roads contribute a great deal to pollution and higher consumption of fuel.

ii) Engine stop

Engine should be switched off if the stop is for more than two minutes particularly at traffic junction or crossings. This will save fuel and reduce the emission.

iii) Educational programme

Educational and awareness programmes should be organized to educate and train the drivers and owners about traffic system.

iv) Transport infrastructure

Transport infrastructure should be improved. Pressure on road traffic can be reduced by diverting alternative transport system *viz.* (a) local trains (b) -water based transport system (c) cycle tonga/horse cart.

(v) Zoning strategy

(a) Setting up of source specific and area wise air quality standards and time bound plans to prevent and control pollution.

(b) Proper localization of industries

(c) Location of industries as per environmental guidelines

(vi) Laws

(a) Enforcement of air pollution control norms

(b) Introduction of environmental audit

(c) Environmental impact assessment of cities at planning stage

(4) Moral And Educational Ways

- » Common man should be made aware of pollution and related problems
- » The public pooled transport system should be encouraged.
- » People should be discouraged for using two wheelers, Car but be encouraged to use bicycles.
- » Environmental awareness programmes should be prepared on how to save environment and energy by involving occupants, NGO's residents, intellectuals scientists, economists and medical supervisor.
- » Environmental tax should be imposed on polluting bodies. This will generate funds which can be utilized strictly in getting new technology to researchers and Non –Government Organisation (NGOs) *etc.*

Chapter - 10

Study Plan for Air Pollution Monitoring and Impact Analysis

Urban Air Pollution and Its Impact on Roadside Plants

Urban Air Pollution

Modern industrial society produces a large number of gases and particles in quantities to harm human, plant or animal life. The amount of pollutants released to the atmosphere by fixed or mobile anthropogenic sources is generally associated with the levels of economic activity. Deterioration of air quality is a major environmental problem in many urban centers in both developed and developing countries. Automobiles constitute a major source of air pollution in the urban areas. Motor vehicle emissions result from fuel combustion or evaporation. The most common types of transport fuel are gasoline and diesel. CO_2 and water vapors are the main products of the combustion, emitted in vehicle exhaust. The major pollutants emitted from the vehicles are particulate matter (suspended fine and ultra fine), sulphur dioxide, oxides of nitrogen, carbon monoxide and metals.

Air pollution in cities is adversely affecting trees on one hand but on the other hand trees, vegetation and soil are very effective for trapping and absorbing air pollutants. The critical load of urban trees can be regarded as the accumulated amount of a pollutant, which will result in physical damage. Trees and woodlands play an important role in the improvement of urban air quality because of its capacity to act as efficient interceptions of airborne matter, therefore plants are widely used as passive bio-monitors in urban environment.

Sources of Urban Air Pollution- A Case Study of Lucknow City

Population- Continuous rise in population is the root cause of air pollution in cities.

Transportation- It includes

» Rise in vehicle Population
» Poor condition of vehicles
» Poor vehicle design
» Predominance of older vehicles
» Predominance of 2 stroke two wheelers
» Poor maintenance of vehicles
» Absence of catalytic converter in vehicle

Fuel- It includes:

» Type of fuel
» Fuel combustion
» Fuel quality

Road condition. It includes

» Damaged roads
» Narrow roads

Traffic management- It includes

» Mixed vehicular traffic
» High vehicle density
» Improper traffic management system

AIM'S AND OBJECTIVES

» To study ambient air pollution level
» To identify the affected area due to motor exhaust.
» To study the effect of air pollution on selected bio-chemical parameters in different plant species to see the tolerance effect.
» To evaluate the accumulation of heavy metals in different plant species.

Study Area - Lucknow At A Glance

Geographical position	-	26°50'53.6 N Latitude 80°5638.1 E Longitude
Above mean sea level	-	120 m
Area	-	528 sq km
Population	-	3.68 millions (33.25 % rise)
Road Transportation	-	Two wheelers- 79.61%
Three wheelers	-	2.26%
Four wheelers	-	10.69%
Bus	-	0.52%
Passenger vehicle	-	0.97%
Other	-	5%
Total Vehicle Population	-	422188
Growth from last year	-	7.65%
Total No. of Petrol Pumps	-	82
Motor Spirit Consumption	-	51911KL (20.1%)
Diesel consumption	-	138623KL (13.62%)
Major source of pollution	-	Vehicles
Monitoring Locations		
Residential	-	Gomti Nagar, Indira Nagar, Aliganj and Vikas Nagar
Commercial	-	Hazratganj, Hussainganj, Charbagh, Alambagh, Aminabad & Chowk
Industrial	-	Amausi & Talkatora

Meteorology

Season	Pressure (mmHg)		Temp (° C)		RH (%)		Rainfall (mm)	WS (Km/Hr)
	8:30	17:30	Max	Min	8:30	17:30		
Summer	994.63	990.9	37.5	21.5	44.7	26	9.1	
Monsoon	987.2	983.3	35.1	26.9	76	66.6	231.7	4
Post monsoon	997.9	994.7	31.7	19.2	75	63	74.47	2.1
Winter	1033.7	1000.7	24.8	9.8	77.7	52	15.6	2.3

Monitoring Schedule

Air

In the present study 24 hour sampling was done using high volume sampler (HVS) and respirable dust sampler (RDS). At each location monitoring was carried out for 2 days per month for the year 1999 to 2000 to cover all the four seasons.

Plant and Soil

Sampling was carried out during April representing summer, July representing monsoon and December representing winter at all the locations. Leaf samples were collected from all the sides of the plant (top, middle and bottom of the canopy) as per method suggested by Rao (1971) and Agarwal and Tiwari, (1996)

Air Sampling Schedule

Air pollutants were monitored according to the schedule given in table

Particulars	TSPM and RSPM	SO_2	NO_x
Sampling equipment	High volume sampler and Respirable Dust Sampler	High volume sampler	High volume sampler
Flow rate	1.2-1.4 cum/min	0.50.1 lit/min	0.50.1 lit/min
Duration of sampling	24 hours	24 hours	24 hours
Frequency	24 hourly	8 hourly	8 hourly

MATERIALS AND METHODS

Air

Parameter	Instrument used	Reference of Methodology	Method of Analysis
Particulate Matter	Respirable dust sampler	IS 5182 (Part 4), 1999	Gravimetric
SO_2	RDS equipped with gaseous attachment	IS 5182 (Part 2), 2001	Colorimetric
NO_x	RDS equipped with gaseous attachment	IS 5182 (Part 6), 1975 reaffirmed in 1998	Colorimetric

Biochemical Parameters

Parameter	Reference of methodology
Chlorophyll	Arnon 1949, Strain *et. al.* 1971
Phaeophytin	Vernon *et. al.* 1960
Carotenoid	Duxbury *et. al.* 1956
Protein	Lowry *et. al.* 1951
Peroxidase	Srivastava *et. al.* 1972, Angelini and Fedrico, 1989

Metals

Metals in ambient air, plant leaves and in surface soil were analyzed by acid digestion method using Atomic Absorption Spectroscopy.

Metals selected for analysis were

(i) Lead (Pb)
(ii) Manganese (Mn)
(iii) Iron (Fe)
(iv) Copper (Cu)
(v) Chromium (Cr)
(vi) Zinc (Zn)
(vii) Cadmium (Cd)

Plants

» Roadside plants were selected for monitoring and impact analysis.
» Selected species were:

Botanical Name	Common Name
Azadirachata indica	Neem
Ficus religiosa	Peepal
Polyalthia longifolia	Ashok
Delonix regia	Gulmohar
Cassia fistula	Amaltas
Mangifera indica	Mango

Biochemical Parameters selected for analysis were:

» Pigment

» Chlorophyll content

» Phaeophytin content

» Carotenoid content

» Protein content

» Peroxidase activity (enzyme)

AMBIENT AIR QUALITY

Analysis data at different locations at Lucknow city showed that all values are on higher side and beyond the NAAQS, in case of TSPM and RSPM which are 200 and 100 mg/m^3 for residential and commercial area respectively. In case of industrial area the concentration of TSPM was within the NAAQS limits which is 500mg/m^3, whereas in case of RSPM values crossed the permissible limits in summer, post monsoon and in winter season.

It is obvious that higher level of TSPM and RSPM in the ambient air was due to automobile exhaust. Results also indicate that higher level was found in the area where automobile transportation was high. It has been found that TSPM and RSPM are significantly influenced by the motor vehicle density and emissions.

Analysis data reveals that TSPM and RSPM level was exponentially higher during winter and summer season. Overall higher level of TSPM and RSPM at all the locations is of great concern due to the effect on human beings. An epidemiological study (Schwartz *et. al.*, 1996, Dockey and Pope, 1996, Donaldson and Macnee, 2001) suggested that atmospheric particulate matter in urban area has a clear correlation with the number of daily deaths and hospitalizations as a consequence of pulmonary and cardiac disease responses. These studies show that measurements of thoracic and alveolar particles (PM$<10\mu$ and $<2.5\mu$ respectively) correlate better with morbidity and mortality, fine and ultra fine particles (RSPM) are released mainly from the automobile exhaust. Diesel fuel is more responsible for fine particles.

At present it has been found that diesel driven vehicles are gradually increasing in the urban area due to cheaper fuel cost. The fine and ultra fine particle fraction also contains organic and inorganic (metals) and other carboniferous particles, which also affect the plant growth, and also affect the biochemical parameters. These particles may also be the source of metal accumulation by plants directly from the air (Dollard, 1986).

In case of SO_2 and NO_X, all the values are within the NAAQS. The concentration of SO_2 and NO_X was greatly influenced by the automobile emissions. Higher concentration was found in commercial areas. Concentration of NO_X was found higher than SO_2. It has been reported that diesel exhaust is mainly responsible for higher values of NO_X. Seasonal variation of SO_2 and NO_X concentration was well marked, higher values were found in summer and winter for NO_X than post monsoon and least in monsoon. In case of NO_X comparatively higher concentration was found in winter season, it may be due to the higher nitrate formation in cold season (Querol *et. al.*, 2001).

It has been found that particulates generated from automobile exhaust especially from diesel engine vehicles are of great concern from pollution point of view. It has also been reported that diesel exhaust particles contain carbon nuclei which absorb a number of organic compounds, which are secondary pollutants, such as polycyclic aromatic hydrocarbons, nitro aromatic hydrocarbons, hetrocyclics quinones, aldehydes and aliphatic hydrocarbons, trace amount of heavy metals, benzo(a) pyrene and 2,3,7,8-tetra chlorodibenzo-p-dioxin.

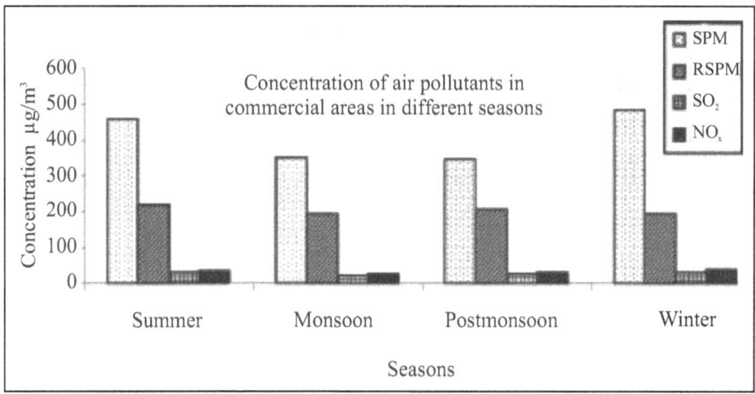

Note : All the values on the graph are the annual and seasonal average.

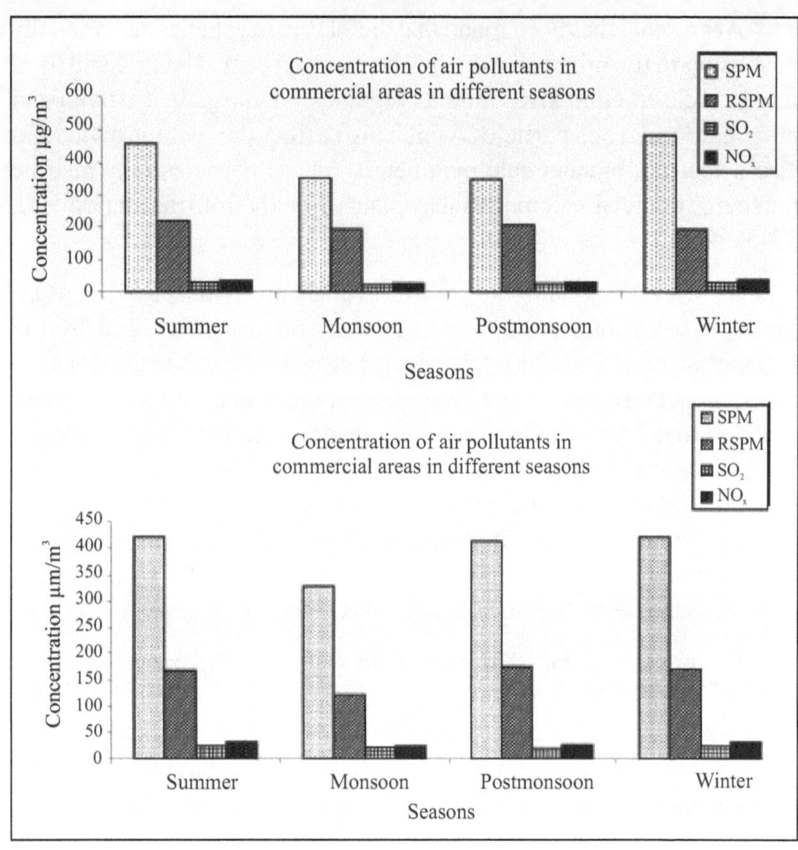

Note : All the values on the graph are the annual and seasonal average.

Air Pollution and Response of Plants to Air Pollutants (TSPM, SO_2 and NO_x) with Respect to Specific Biochemical Changes

The response of plants to air pollution at physiological and biochemical level can be understood by the analysis of biochemical parameters like pigment content (Chlorophyll, Phaeophytin and Carotenoid), protein and peroxidase enzyme.

Chlorophyll Content

Degradation of photosynthetic pigment has been widely used as an indicator of air pollution (Schulze, 1986, Ninave *et. al.*, 2001). Darall (1986) reported that SO_2 might reduce plant growth even at concentration (low), which is not enough to cause visible injury, mainly through its adverse effect on photosynthesis. Chlorophyll content of plant signifies its photosynthetic activity as well as the growth and development of biomass. It is well evident that chlorophyll content of plants varies species to species; age of leaf and also with the pollution level as well as with other biotic and abiotic conditions (Katiyar and Dubey, 2001). Reports of decreased chlorophyll concentration in plant fumigated with SO_2 are often associated with degradation of chlorophyll (Khan *et. al.*, 1990).

Present study revealed that chlorophyll content in almost all the plants varies with the pollution status of the area *i.e.* higher the pollution level lower the chlorophyll content. Several other studies (Khan *et. al.*, 1990 and Agarwal and Tiwari, 1996) also suggest that high level of pollution decreases chlorophyll content in higher plants, particularly the dust fall on leaf surface as well as high concentration of SO_2 and NO_X (Panigrahi *et. al.*, 1992; Kumar and Jayshree, 1999 and Tiwari and Bansal, 1993a). High level of SPM, SO_2 and NO_X at commercial locations might be the cause of low chlorophyll content in all the plant samples at commercial area.

In all the species chlorophyll content was higher in monsoon season, it might be due to the washout of dust particles (which will increase photosynthetic activity) from the leaf surface, low level of pollution and water content of soil as suggested by Rao, 1989. Highest chlorophyll content was found in monsoon then in summer and least in winter season, it might be due to the high NO_X concentration which increase the sensitivity of plants to winter stress as observed by Imiwari *et. al.* (1992). Low chlorophyll content in winter season might be due to the high pollution level, temperature stress, low sunlight intensity and short photoperiod.

Chlorophyll decreases with increasing pollution and also varies species to species. A decrease in chlorophyll content was observed at commercial locations in comparison to residential locations. Amaltas and Ashok (mild) showed a increase in chlorophyll content at polluted areas. On the basis of chlorophyll content, it was found that Amaltas (*Cassia fistula*) and Ashok(*Polyalthia longifolia*), species were less affected due to air pollution hence can be considered resistant species as suggested by Beg, 1990b. Ninave (2001) made similar observations in Nagpur city during a vegetation survey.

It is well known that air pollutants especially SO_2 destroy chlorophyll content in plant and replace Mg^+ of chlorophyll molecule by two atoms of hydrogen forming Phaeophytin and changing the light spectrum characteristics of the chlorophyll molecule and reducing the photosynthetic activity (Voet and Voet, 1990).

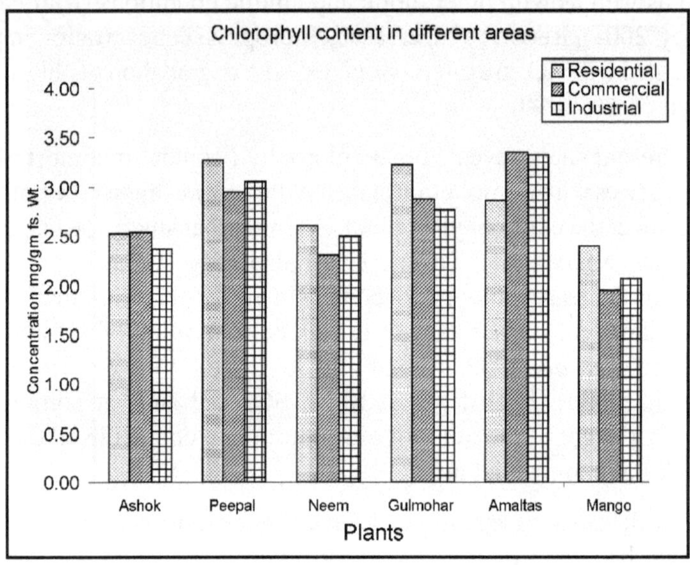

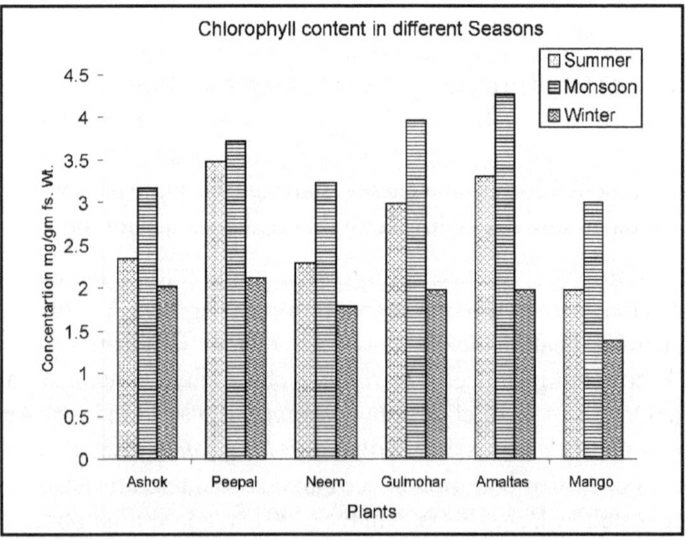

Note: All the values on the graph are the annual and seasonal average.

Phaeophytin Content

Earlier paragraph discussed the analysis results of chlorophyll in different plant species. The trend of Phaeophytin contents of plant leaves was found to be opposite that of chlorophyll content. It might me due to the breakdown of chlorophyll to phaeophytin as suggested by Rao and Le Blanc, 1966. The results indicate that the Phaeophytin content was higher where the chlorophyll content was low.

Higher level of Phaeophytin reduction was found in winter season in all the plant species, when the pollution level was high. The average Phaeophytin content in plant leaves at all the locations showed that values were higher than chlorophyll content, which indicates the conversion of chlorophyll to Phaeophytin or reduced biosynthesis. The highest Phaeophytin was found in Amaltas.

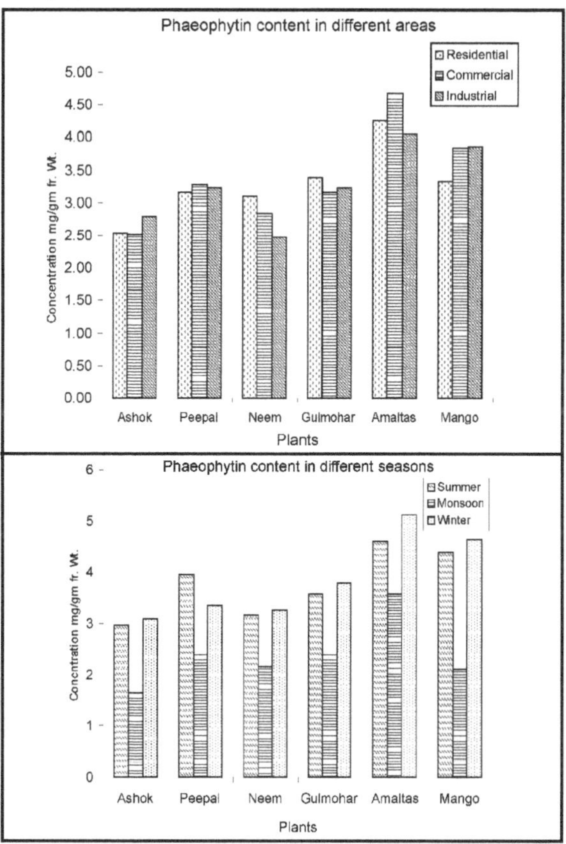

Note: All the values on the graph are the annual and seasonal average.

Carotenoid Content

Carotenoid is an accessory pigment responsible for yellow to orange color in plants and is a free radical scavenger, therefore it is responsible for the protection capacity of the plants (Rawn, 1989). It was generally observed that at polluted sites senescence occurs several weeks early. This early senescence was due to the changes observed in a major group of plant pigment carotenoid (Goodwin, 1958).

It was found that comparatively carotenoid content was decreased in all the species where the pollution level was high along with high vehicular density.. Several other studies by Bhattacharjee *et. al.*, 1994; Pcssarakli, 1999; Sinha *et. al.*, 2002 suggest that higher particulates, SO_2 , NO_x , and metal concentration in soil as well as in air are responsible for low level of carotenoid . Present study also reveals that carotenoid content was low during winter reason and may be due to the breakdown of carotenoid under high pollution level.

As observed by health (1989) the carotenoid content decreases at commercial locations in comparison to residential locations where the pollution level was low comparatively. Amaltas showed no changes and mango showed an increase in carotenoid content, hence are least affected by pollution with respect to carotenoid changes.

Base line value of carotenoid content was highest in Amaltas indicating that it is better equipped with anti oxidative system. Study shows that under pollution stress defense system of plants gets activated to reduce the burden of pollution.

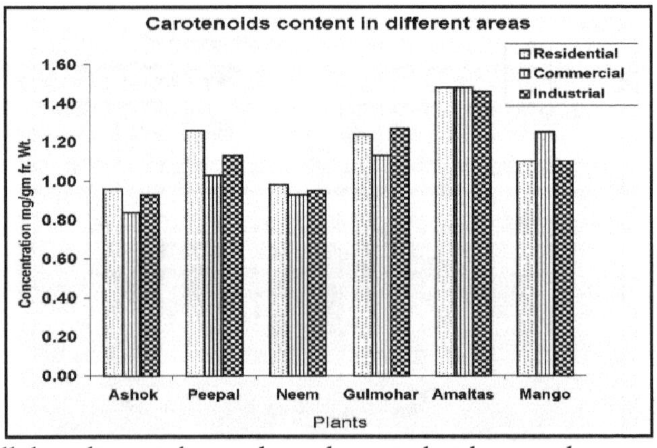

Note: All the values on the graph are the annual and seasonal average.

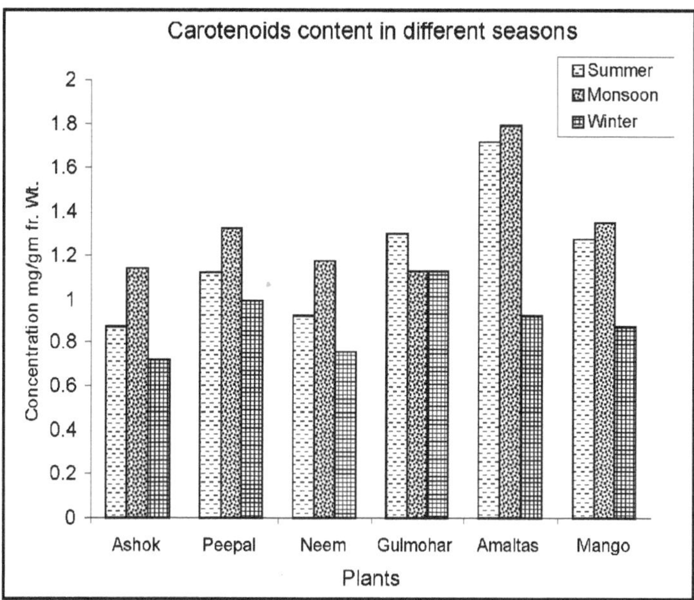

Note: All the values on the graph are the annual and seasonal average.

Protein Content

Protein carries out important enzymatic activities and is vital for the rapid rate of biochemical reaction in cells. Two other important functions of proteins are, as major H^+ ion buffers and as structural component of cells. Many have reported decrease in protein content on SO_2 fumigation (Khan *et. al.*, 1990; Nandi *et. al.*, 1990) and the tentative reasons attributed to this loss are either the breakdown of the existing protein molecules or reduced biosynthesis.

The protein content was found to reduce in all the plants species, and the maximum reduction was found in plants at commercial locations where the pollution level was comparatively high. It might be due to the high concentration of SO_2, NO_X and particulates at commercial locations as suggested by Nandi *et. al.*, 1990 and Rao and Dubey, 1990.

The effect of air pollutants depends on the physical and chemical properties of the pollutants and also with the receptors (Krupa, 1996). Other studies reported that SO_3^{-2} and HSO_3^- generated by dissolution of SO_2 in cytoplasm are detoxified by oxidation to less toxic sulphates, in this process reactive radicals are formed (Asades & Kiso, 1993,

Tanaka *et. al.,* 1985), which are more dangerous than sulfurous compounds (Ziegler, 1975). The reactive radicals bring about changes in Chlorophyll and protein content (Pierre and Quernoz, 1981; Melhorn *et. al.,* 1980).

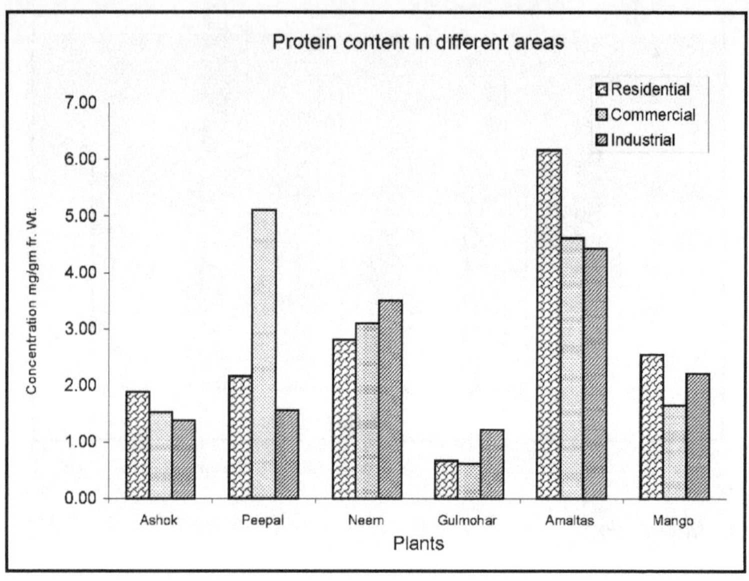

Rabe and Kreeb (1974) studied enzyme activities, chlorophyll and protein content in plants as indicators of air pollution and mentioned the decrease of chlorophyll and protein content.

Results revealed that protein content was higher in monsoon season at all locations in all the plant species. During Monsoon the level of pollutants were comparatively low than summer and winter. Comparatively protein content was low in all the plant species at the locations where the vehicular density was high. It indicates that the air pollutants emitted from the automobile exhaust are responsible for the reduction of protein levels. Kumar and Jayshree, 1999, made similar observations.

It is also remarkable that lower protein content of leaf was found in the commercial areas with higher traffic density where the pollution level was high. It is well known that SO_2 and NO_X in combination adversely affect the plants as well as the protein content as observed by Amundson *et.al.*, 1986.

Among the plant species studied, comparatively higher levels of protein content was found in Neem and Amaltas in all the seasons. *The results indicated that at same environmental pollution levels, pollutants are less effective for Neem and Amaltas. High protein content shows the better protection capacity of plants.* Protein dissociates to form amino acids and the energy is utilized for routine metabolic activities.

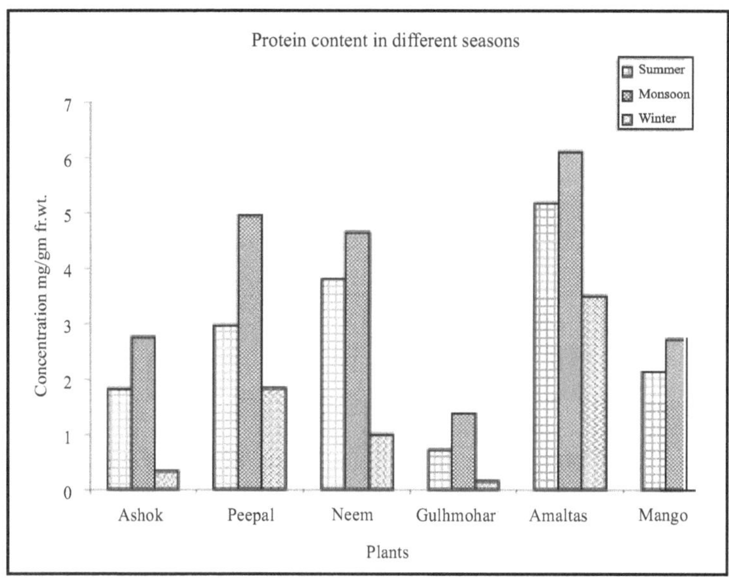

Note: All the values on the graph are the annual and seasonal average.

A decrease in protein content was observed at commercial location in comparison to residential locations. Such a decrease could be attributed to either breakdown of existing protein or their reduced biosynthesis as suggested by Malhotra and Khan, 1984. Increased activity was observed in Neem and Peepal at commercial locations showing the utilization of sulphur and nitrogen from environment for their nutrition.

Peroxidase Activity

Peroxidase (POD) occurs in a wide variety of plants including deciduous and evergreen. The level of peroxidase enzyme is universally accepted as an indicator parameter to environmental pollution especially air pollution (Nandi *et. al.*, 1990). Studies of Karjalainen *et. al.*, 1992 and Khan *et. al.*, 1990, suggest that Peroxidase is a specific indication of SO_2 and NO_X pollution. In combination SO_2 and NO_x induced strong Peroxidase activity in plants. Several other factors were also involved in the enhancement of POD activity like heavy metals as observed by Grunhagel *et. al.* (1981). Some times it has been found that micro and macronutrient essential for plants induced POD activity.

Present study revealed that the higher peroxidase activity was found in plant leaves of Ashok, Peepal, Neem, Gulmohar and Mango in areas with high traffic movement. Similar observations were made by Bragaloni (1995) with different species. Amaltas showed decreased activity at commerciall locations. Results also indicate that activity was higher in winter when the pollution level was high followed by summer and least in monsoon when the pollution level was comparatively low, the same trend was observed by Puccinelli *et. al.* (1998) with different species.

All the six species of plants studied reacted to air pollution level in Lucknow city. POD activity varies species to species, season to season and with the level of air pollutants in ambient air. It was also found that deciduous plants species with broad leaves (Peepal) showed higher POD activity. Puccinelli *et. al.* (1998) also observed high POD activity in deciduous plants. Evergreen plants like Ashok and Mango also showed high POD activity at commercial areas with high SPM, SO_2 and NO_x concentration, in winter season.

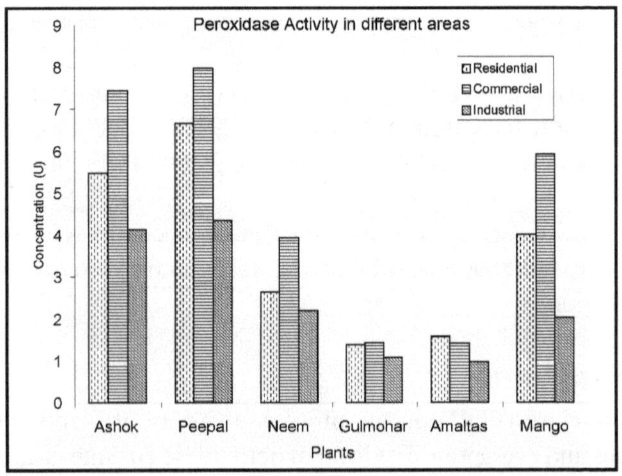

Note: All the values on the graph are the annual and seasonal average.

On the basis of POD activity, it can be suggested that Ashok, Peepal and Mango are highly sensitive to the air pollution whereas Neem Gulmohar and Amaltas were less sensitive and showed low POD activity. Over all POd activity was lowest in Amalts i.e. low accumulation of H_2O_2 and lesser generation of free radicals.

An increase in POD activity was observed at commercial locations in comparison to residential locations. Ashok and Neem showed maximum rise, hence are sensitive to air pollution as suggested by Bragaloni (1993) in other species.

Present study has shown that at commercial locations and in winter season, stress on plants was maximum. As a result decrease in chlorophyll and protein content was maximum at commercial locations in comparison to residential and industrial locations. A decrease in the protection capacity of plants in terms of carotenoid content was maximum at commercial area. A decrease in protein content was also observed at commercial areas. A high POD activity was observed at the commercial locations due to intense vehicular traffic. Biochemical changes in leaf tissue are a response stress (Pollution, heat/ summer and cold/winter). Biochemical changes in plant follows the air quality trends hence biochemical parameters seems to be suitable marker of air pollution and can be used to draw or edit air quality maps.

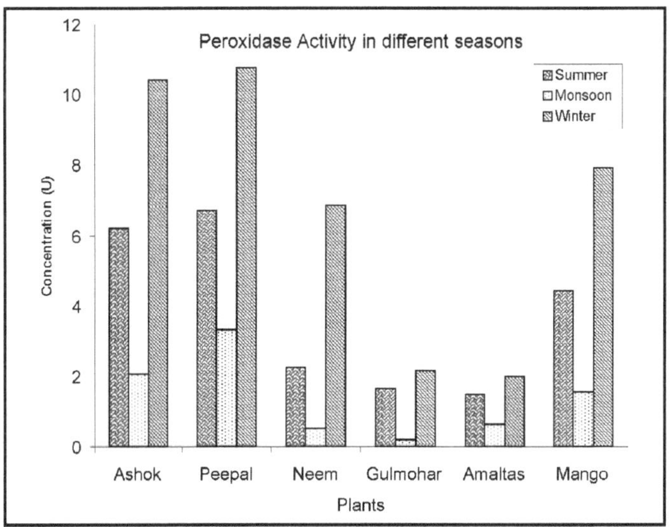

Note: All the values on the graph are the annual and seasonal average.

Heavy Metal Content in Urban Air and Soil and Its Accumulation in Roadside Tress

Heavy metal contamination of urban air, soil and vegetation related to traffic in industrial countries has become a serious environmental problem. Tetraethyl lead is still added in some countries to gasoline as antiknock agent, and Zn, Cu and Cd are components of some alloys, pipes, wires, tyres and lubricating oil. It has been found that lead and unleaded petrol and diesel exhausts emit heavy metals into the roadside environment as a result of combustion and mechanical abrasion and wear (Lagerwerff and Specht 1970, Seaward and Richardson, 1990, Xiong 1998, Monacci and Bargali, 1997, Caselles *et. al* 2001).

Metals in Air (Suspended Particulate Matter)

In our study area, the major source of metal pollution in the ambient air is vehicular exhaust. Mainly two types of fuels are used by vehicles i.e. petrol and diesel. Combustion of these fuels releases a number of complex organic and inorganic pollutants including metals. Recently particulate matter from vehicular exhaust is categorized in different fragments like PM, PM_{10}, $PM_{2.5}$ and PM_1 depending on the size.

Combustion of fuel especially the diesel is one of the major source of fine and ultra fine particles (<10μm) in the urban area. It has been reported that organic, elemental (metal) and mineral carbon fractions accounted for 70%, 26% and 5% of total carbon In PM_{10} respectively (Querol *et. al*, 2001).

In our study it has been found that heavy metal like Pb, Mn, Fe, Cd, Cr, Zn and Cu are present in ambient air in different degrees. In case of lead higher concentration was found in the commercial and high traffic areas namely Hazratganj and Alambagh during winter season, The levels were beyond the permissible limits of CPCB standards. Concentration of lead in different places is mainly dependent on traffic volume of the specific area *(description of the area was given in chapter 1)*.

For other metal study, the metal concentration was higher in commercial areas with heavy traffic volume than industrial and residential areas. Considering all the sides it is revealed that minimum concentration was found in Gomti nagar, which is residential area with less traffic movement and can be considered as control area.

It was also revealed that concentration of metals was higher in winter season, it may be due to the lower winter dispersive condition of auto exhausts.

Metals (annual average) in different areas of the city (µg/g d.w.)

	Residential	Commercial	Industrial
Pb	0.1	0.41	0.66
Mn	0.07	0.14	0.12
Fe	1.93	4.2	3.82
Cu	0.55	0.2	0.1
Cr	0.02	0.03	0.03
Zn	0.53	1.09	0.53

Seasonal variation of metals in ambient air

	Summer	Monsoon	Postmonsoon	Winter
Pb	0.28	0.13	0.23	0.75
Mn	0.12	0.05	0.08	0.19
Fe	3.59	1.91	2.7	5.31
Cu	0.36	0.05	0.12	0.67
Cr	0.02	0.01	0.02	0.06
Zn	0.91	0.45	0.68	1.14

Metals in Soil

Concentration of metals in roadside soil shows a great variation, depending on the type of area and traffic volume. Normally results revealed that the higher concentration of metals was found in commercial area with higher automobile transportation. Generally in all the cases metal concentration was found high in winter and next in summer season.

Obviously the higher levels of metals along the road are attributable to the traffic movements. Major source of metal accumulation in roadside soil is due to the fall of particulates released from the auto exhaust. It is remarkable that during the monsoon season thc concentration of metal was found at lower level and may be due to the washout of the upper soil by rainwater run off. In each case the concentration of lead was in higher level it may be due to the partial use of leaded petrol. Mn and Cr showed a little variation with the change of season and Fe, Cu, Zn and Cd were found in higher concentration in commercial areas.

From the results it may be concluded that concentration of metals like Pb, Fe, Cu, Zn and Cd mainly depends on the automobile pollution *i.e.* deposition of automobile carbon soot on the surface soil.

Metals (annual average) in different areas of the city (µg/g d.w.)

	Residential	Commercial	Industrial
Pb	67.66	152.90	106.90
Mn	204.68	260.16	212.46
Fe	35470.47	10894.49	12920.70
Cu	138.02	327.02	167.39
Cr	6.15	9.74	6.82
Zn	68.26	98.70	64.49
Cd	0.20	0.35	0.24

Seasonal variation of metals in roadside soil

	Summer	Monsoon	Postmonsoon	Winter
Pb	109.04	113.26	107.76	126.63
Mn	256.24	232.91	233.35	212.36
Fe	9645.41	12120.05	11045.75	10628.76
Cu	269.06	207.12	235.51	238.00
Cr	8.92	6.68	8.28	8.33
Zn	74.08	68.44	93.96	96.74
Cd	0.50	0.20	0.18	0.24

Summer

	Residential	Commercial	Industrial
Pb	19.41	25.67	27.52
Mn	6.68	7.47	5.34
Fe	82.06	96.4	105.51
Cu	6.08	8.82	8.82
Cr	0.48	0.37	0.81
Zn	9.62	11.93	8.25
Cd	0.35	0.55	0.92

Monsoon

	Residential	Commercial	Industrial
Pb	21.05	21.11	20.4
Mn	15.23	16.19	13.34
Fe	106.87	116.18	122.23
Cu	10.75	12.85	12.66
Cr	0.08	0.13	0.11
Zn	16.53	21.57	21.88
Cd	0.08	0.14	0.15

Winter

	Residential	Commercial	Industrial
Pb	13.78	15.38	15.85
Mn	14.87	15.94	14.93
Fe	107.9	111.26	110.49
Cu	9.09	12.85	11.42
Cr	0.09	0.11	0.11
Zn	14.75	16.81	14.27
Cd	0.07	0.16	0.1

Accumulation of different metals by different plants was as follows:

Pb- Amaltas>Gulmohar>Ashok>Neem>Mango>Peepal
Mn- Amaltas>Neem>Peepal>Mango>Gulmohar>Ashok
Fe- Neem> Amaltas>Ashok> Mango>Peepal>Gulmohar
Cd- Mango>Amaltas>Ashok>Neem> Peepal> Gulmohar
Cr- Neem>Peepal>Ashok>Amaltas>Mango>Gulmohar
Zn- Amaltas>Neem>Ashok>Mango>Peepal>Gulmohar
Cu- Gulmahar> Amaltas>Neem>Peepal>Ashok>Mango

Plant leaves are an important part to monitor the pollution effects on plants. There are three aspects to monitor the pollution effects on plants

i. Phenotypic symptoms (visible symptoms)
ii. Biochemical changes
iii. Metal concentration in leaf tissue

In our present study we have estimated the metal concentration in plant tissue (leaves). Concentration of metals indicated the pollution status around the area. Plants normally uptake or accumulate metals via root which is available in the soil solution. Metals from the automobile exhaust ultimately fall on the soil surface, which is normally inorganic. A number of factors are responsible for the uptake of metals from the soil by roots and also translocation to the upper parts. Some factors such as soil pH, soil texture, moisture, bio-available concentration of metals, organic matter and also plant species including different parts of plants.

Lead is a toxic element and not required by plants. At elevated levels Pb is toxic to any living being. Lead enters in the food chain mainly via plants. The critical concentration of lead varies species to species. In our study main source of lead was automobile exhaust. During our study period partially leaded petrol was used across the city.

Results revealed that lead concentration was high during summer season in all the plant species and comparatively less in winter season, it might be due to the transportation activity and other abiotic factors. Normally the lead uptake by root remains in the lower part of the plant. Lead concentration is higher than the normal background concentration, which is 2.1 and 2.5 mg/kg in grass and clover plants (Pais and Jones, 1997). The phenotypic appearance of the plants did not show any effects.

Among the plant species, little difference was found in particular season. In all the seasons Amaltas showed highest average concentration 27.72 µg/g d.w. during summer, which indicate the higher capacity at Pb accumulation. Low concentration was found in mango leaves in monsoon and winter season, it indicated less accumulation. Generally it has been found that most plant species are sensitive to lead tissue concentration between 25-85 µg/g d.w. (Pahlsson, 1989).

It is not known that Mn is emitted from automobile exhaust. Mn is an essential element and is required for many biological activities specially chlorophyll pigmentation. Comparatively Mn concentration was higher in heavy traffic area. In general hetrogenous accumulation was found in all seasons. Overall Amaltas species showed higher concentration of Mn in their leaves in all seasons.

Iron is an essential nutrient as it actively participates in different metabolic pathways. Fe is normally accumulated in roots. The concentration of Fe in leaf tissue showed little variation among different locations. Results revealed that higher accumulation was found in commercial areas like Charbagh, Alambagh, Aminabad and industrial area like Talkator where the pollution level was higher than other locations and indicates that source of Fe might be automobile exhaust. Overall the Fe accumulation in plant tissue showed heterogeneous accumulation. Comparatively higher accumulation was found in monsoon and winter season

Cadmium is not required by plants. At higher concentration Cd is toxic to both plants and animals. In plants Cd ions have a strong affinity to the sulphahydral and phosphate groups of certain compounds that are involved in metabolic processes. This affinity might be the reason of Cd toxicity (Carlson and Bazaz, 1977).

Cd inhibits photosynthesis and transpiration in plants. The results indicated that heterogeneous accumulation was found in all the plants.. Little variation was found at different locations. Accumulation of Cd during summer was comparatively less , in some cases Cd was not at detectable level.

Overall concentration of Cd at all the locations in all the plant samples was within the normal range as suggested by the Bowen (1979). Which is in the range of 0.1 to 2.4 μg/g d.w. in case of terrestrial plants.

It has been established that Cr is an essential nutrient (Merts, 1969) but not for plants (Huffman and Alloway, 1973). Normally the Cr has been found in plant tissue in small quantity. The toxicity of Cr at high concentration has been reported in different plant species (Hunter and Vergano, 1953, Joshi *et. al*, 1999).

Results indicate that heterogeneous accumulation was found at all the locations among the six plant species. Different variation was found in different seasons. But in summer season concentration of Cr was below detection limit at certain locations. Comparatively higher level of Cr was found in locations where automobile pollution was high like Hazratganj, Husainganj, Charbagh *etc.*

Zn is an important element for both plants and animals. It plays an important role in several plant metabolic processes, it activates enzymes, is incorporated in metallo-enzymes of the electron transport system and is involved in protein synthesis, carbohydrate, nucleic acid and lipid metabolism. It forms complexes with RNA, DNA and affects their stability.

The results revealed that the accumulation of Zn in all the plant species was heterogeneous and comparatively higher accumulation was found in monsoon season then in winter and least in summer.

On the other hand the concentration of Zn was higher in the commercial area with heavy traffic movement. Its is well proved that the source of Zn in urban area is mainly from automobile exhaust. Zn concentration was high in soil, air and plant tissue in traffic areas. Generally plants are sensitive to leaf tissue concentration ranging from 60-90 μg/g d.w. (Macmical and Beckett, 1985). But our results showed much lower concentration in leaf tissue in all plant species. It is also reported that most plants show potential phytotoxicity at leaf tissue concentration above 200 mg/kg d.w. (Davis and Beckett, 1977).

Copper is an essential element for plants because it plays an important role in carbohydrate, nitrogen and cell wall metabolism. It acts as a co-enzyme and is involved in seed production and disease resistance. At excess concentrations, however Cu ions inhibit photosynthesis and respiration thereby decreasing chlorophyll content.

Comparatively the accumulation of Cu concentration in plant tissue in all the plant species was found higher in commercial areas like Hazratganj, Charbagh. Lower concentration was found in summer season and high level was found in monsoon season. Comparative less marked difference was observed in seasonal accumulation.

In most cases concentration of Cu in leaf tissue was within the range of normal tissue concentration in terrestrial plants grown in unpolluted soils which varies from 5-15 μg/g d.w. (Bowen, 1979), and the toxicity symptoms are visible when the plant tissue concentration of Cu reaches 20-30 mg/kg d.w. (Frank *et. al*, 1986; Davis and Beckett, 1970; Reuther and Labanauskas, 1966). Generally Cu is immobile in soil and remains at the surface layer and is primarily associated with soil organic matter.

Conclusion

Particulate Matter- Study found that SPM and RSPM were significantly higher in ambient air of the Lucknow city.

Sulphur Di-Oxide and Oxides of Nitrogen- Study shows that SO_2 and NO_x concentration at all the locations were within the prescribed standards.

Oxides of Nitrogen-In the study it was found that NO_x concentration was comparatively higher than SO_2.

Seasonal Variation- Level of pollutants was high in winter and least in monsoon season.

Biochemical Changes- It was observed that Air pollutants, metallic content in air and soil affect biochemical parameters in urban woodlands.

Chlorophyll Content- Decreases with increasing pollution. Ashok, Amaltas and Peepal were found less affected with respect to chlorophyll changes.

Phaeophytin Content- Higher phaeophytin reduction was found in winter season. Present study showed that Amaltas, Peepal and Mango were sensitive to air pollution with respect to Phaeophytin content in their leaves

Carotenoid content- It decreases at commercial locations. Amaltas, Mango and Neem show minimum reduction in carotenoid content.

Protein content- Decrease in protein content is observed at areas with higher vehicular movement.

Peroxidase activity- POD activity increases under pollution stress

Metals

- » High metal content in ambient and soil was observed at commercial locations.
- » Among the toxic metals (Pb, Cd and Cr), concentration of Pb was highest in ambient air.
- » Amaltas can be planted in areas with high Pb, Mn and Zn content in air and soil.

- » Neem can be planted in areas with high Fe and Cr content in air and soil.
- » Gulmohar can be planted in areas with high Cu content in air and soil.

Green belt development

Since chlorophyll is the main component responsible for the vegetational growth of the plant, any decline in its content will adversely affect the growth of plant. So the plants which were resistant to air pollution with respect to chlorophyll changes can be the basis for the selection of tree species for developing green belt. Ashok and Amaltas were least affected by air pollution decrease in chlorophyll content.

In view of the above this type of study would be helpful in finding pollution sensitive and tolerant species. Sensitiveness and tolerance varies species to species. It not only depends on concentration of pollutants in ambient air but also on the climatic conditions of the area.

Results of the study can be utilized under similar geographical, geological, metrological, environmental (population, vehicle and fuel status) and biological conditions.

Chapter - 11

GREEN BELT DEVELOPMENT

Plants play an important role in maintaining ecological balance by actively participating in the cycling of nutrients and gases like carbon dioxide, oxygen and air pollutants. Sensitivity and responses of plants to air pollutants is variable. Some plants are sensitive and act as bio indicators of air pollution (Kozhouharov *et. al.* 1985, Guderian, 1998). Others which are tolerant may act as sink to air pollutants. However this aspect needs more investigation for their use as indicators of air pollutants and in abatement of air pollution.

Today there is great need to increase public awareness to plant trees in urban areas to enhance the beauty of the concrete landscape and for healthy living. Within past decade there have been many reports on the need for increased tree planting in the urban environment (DoE, 1990). About 10% of UK land area is in urban use (DoE 1995b), within which there are many parks, gardens and open spaces.

Trees and Woodlands warrant a special place in relation to enhancement of the urban environment (Green Spaces Task Force, 2001). Trees and vegetation are effective in trapping and absorbing many pollutant particles and the meteorological conditions are directly correlated with particle in urban environment (Croxford *et. al.*, 1996).

The complexity of the urban environment greatly influences the establishment and management of trees, which are planted and grown there. Federer (1971) described three broad classes of urban ecosystem in which trees are grown. These were:

» Areas with high evaporative or transporative surfaces; *e.g.* parks and wide streets –hot during the day and cold at night;

» Areas open to sky but dry; *e.g.* large squares and car parks-similar microclimate, to previous one but drier; and

» Areas sheltered by buildings ; *e.g.* narrow streets and courtyards-cool during the day and warm at night (*i.e.* small diurnal temperature fluctuations).

As suggested by many workers a great deal of emphasis be placed on the selection of appropriate tree species for their particular urban settings (Thoday,1966) Broadleaves are generally considered to be more hardy of pollution. Plants with broad leaf renew their filtering apparatus, *i.e.* their leaves, each year, therefore decreasing their accumulated annual load of toxic particles. However as the leaves falls these can then accumulate in the soil, causing other physiological damage, particularly if the particles are not readily leached or mobilized from the soil (Wang *et al.,* 1995).Broadmeadow and Freer-Smith (DOE, 1996) have suggested that pollutant –tolerant trees which exhibit high stomatal conductances should be planned in high-pollution 'hot-spots' in order to absorb contaminates and, therefore, improve air quality, an approach also recommended by Good (1990). Broadmeadow and Freer-Smith (1996) show that urban trees could absorb significant amounts of both O_3 and SO_2 (Up to 21 and 20 % of exposure concentrations during episodes of O_3 and SO_2 respectively) both of which are associated with particulate forms of pollution; *i.e.* VOCs and sulphate.

Development of green belt and trees or vegetation has a great importance in urban areas and the critical loads of pollutants can be determined by the accumulated amount of pollutant, which will result in physical damage (*i.e.* threshold for injury) (Farmer, 1995 and Greenhalgh and Worpole, 1995). Moll (1996) states that 12% of air pollution problems in cities are attributable to urban heat island effects. McPherson *et.al.* (1994) estimated that the trees removed approximately 234 tonnes of PM_{10} improving average hourly air quality by 0.4% in Chicago. Similarly Nowak *et. al.* (1997) calculated that the trees of Philadelphia improved air quality by 0.72% in the process of PM_{10} reduction.

Pollution control by trees in Chicago (11% tree cover) in 1991 was estimated at 234 tonnes for particulate matter less than 10 microns, 98 tonnes for nitrogen di oxide and 93 tonnes for sulphur dioxide (Nowak,

1994). The majority of the pollution removal by trees occurs in the leaf in daytime conditions, as this is the time when leaf surfaces are actively transpiring and pollution concentration can reach their maximum. Individual tree size affects total removal per tree. Average pollution removal for individual trees in Chicago in 1991 (for 5 major pollutants) ranged from 0.05 pounds/year for a tree less than 30 inches in trunk diameter to 3.1 pounds/year for a tree greater than 30 inches in trunk diameter (Nowak, 1994).

In 1994 trees in the New York city removed an estimated 1821 metric tonnes of air pollution to an estimated value to society of $9.5 million. Air pollution control by urban forest in New York was greater than in Atlanta (1196t, $6.5) and Baltimore (499t, $2.7) but pollution removal per m^2 of canopy cover was fairly similar among the cities (New York: 13.7g/m^2/yr, Baltimore: 12.2 g/m^2/yr, Atlanta: 10.6 g/m^2/yr). These standardized pollution removal rates differ among cities according to the amount of the air pollution, length of the leaves in the season, precipitation and other meteorological variables. Large healthy trees greater than 77 cm in diameter remove approximately 70 times more air pollution annually (1.4kg/yr) than small healthy trees less than 8cm in diameter (0.02kg/yr). Air quality improvement in New York city due to pollution removal by trees during day time in the leaf season averaged 0.47% for particulate matter, 0.43% for sulphur dioxide and 0.30 for nitrogen dioxide, where there was 100% tree cover short term improvements in air quality for pollution removal by trees was as high as 14%for SO_2, 13% for PM and 8%for NO_2.

Chapter - 12

Sampling and Quantitative Estimation of Major Air Pollutants Biochemical Parameters of Plants and Heavy Metal Analysis

Air

Estimation of Total Suspended Particulate Matter (TSPM) and Respirable Suspended Particulate Matter (RSPM)

Requirements

Instrument

High Volume Sampler (HVS) APM 415-411 (Envirotech Instrument India), Respirable Dust Sampler (RDS) APM 451 (Envirotech Instrument India) with midget impinger containing absorbing solution were used.

Analytical Procedure for HVS and RDS

Sampler is a basic instrument used primarily for measuring concentration of TSPM and RSPM in atmospheric air. Gaseous sampling attachment uses wet chemical methods for the absorption and detection of gaseous pollutants. Atmospheric air was passed through absorbers containing suitable reagents.

Sampler Sequence

A fresh, pre-weighed glass fiber filter paper is installed in filter holder and cyclone cap in the holder in case of RDS. Fluid in the orifice meter is adjusted at zero level. In case both particulate and gaseous pollutants are to be sampled impingers are filled with suitable reagents and placed in ice tray. Impinger outlet is connected to individual inlets of the gas manifold. Initial flow rate of gases is measured.

Prior to completion of the sampling as per schedule, the following were noted.

i. Flow rate indicated by orifice meter and rotameter is recorded.
ii. Final sampling time indicated by the time totalizer is recorded.
iii. Final flow rate of gas is measured.
iv. Filter paper is removed, folded along its length and stored in a clean polythene bag.

Estimation of Air Pollutants

Duration of sampling for gases and suspended particulate matter are 24 hours and the frequency is 8 hourly and 24 hourly respectively.

Calculation

The volume of air at normal temperature and pressure (NTP) (25°C and 760 mmHg) is calculated as follows in the case of TSPM, RSPM and Metals.

$$V_S = (L_1+L_2)/2 \times T \times 60$$

where

VS	=	volume of air in liters at standard conditions
L1	=	Initial flow rate reading
L2	=	Final flow rate reading
T	=	Total time in hours
60	=	Time conversion from hours to minutes

Total Suspended Particulate Matter (SPM) [IS: 5182 Part IV 1987]

Suspended particulate matter is collected on the pre-weighted filter paper EPM 2000 for 24 hours as per sampling schedule at different locations. The volume of air is measured in cubic meters.

Calculation

HVS

TSPM ($\mu g/m^3$) = Difference in weight (gm.) x 106 / Volume of air sampled (m3)

RDS

RSPM ($\mu g/m^3$) = Difference in weight (gm.) x 10^6 / Volume of air sampled (m^3)

Cyclone Cap ($\mu g/m^3$) = Difference in weight (gm.) x 10^6 / Volume of air sampled (m^3)

TSPM ($\mu g/m^3$) = RSPM + Cyclone Cap

Estimation Of Sulphur Dioxide (SO_2) in air [IS:5182, Part-II, 2001 (West and Gaeke)]

Requirements

Reagents

Potassium tetrachloromercurate (TCM)-0.04M

Dissolve 10.86 g mercuric chloride, 0.066 g EDTA, and 6.0 g potassium chloride in distilled water and bring to the mark in a 1 litre volumetric flask. The pH of this reagent shall be approximately 4.0 but it has been shown that there is no appreciable difference in collection efficiency over the range of pH 5 to 3. The absorbing reagent is normally stable for six months, if a precipitate forms, discard the reagent after recovering the mercury.

Caution – highly poisonous if spilled on skin flush off with water immediately.

Sulphamic acid- 0.6 percent

Dissolve 0.6 g of Sulphamic acid in 100 ml of distilled water, prepare fresh when needed.

Formaldehyde—0.2 percent approx.

Dilute 5 ml of formaldehyde solution (36 to 38 percent) to 1 litre with distilled water. Prepare fresh, when needed.

Stock iodine solution 0.1N

Take 12.7 g of iodine in a 250 ml beaker: add 40 g of potassium iodine and 25 ml of water. Stir to dissolve completely, then dilute to 1 litre with distilled water

Iodine solution 0.01N

Prepare approximately 0.01N iodine solution by diluting 50ml of stock solution to 500ml with distilled water.

Starch indicator solution

Titrate 0.4g of soluble starch and 0.002g of mercuric iodide preservative with a little water and add the paste slowly to 200ml boiling water. Continue boiling until the solution is clear, cool and transfer to a glass stoppered bottle.

Stock sodium thiosulphate solution 0.1N

Prepare stock solution by taking 25g of sodium thiosulphate pentahydrate in a beaker add 0.1g of sodium carbonate and dissolve using boiled and cooled distilled water making the solution up to a final volume of 1 litre. Allow the solution to stand one day before standardizing.

To standardize accurately weigh to the nearest 0.1mg, 1.5g of primary standard potassium iodate dried at 180°C, dissolve and dilute to 500ml in a volumetric flask. Take 50ml of iodate solution by pipette into a 500ml iodine flask, add 2g of potassium iodide and 10ml (1:10) hydrochloric acid and stopper the flask. After 5min titrate with stock thiosulphate solution to a pale yellow color. Add 5ml starch indicator solution and continue the titration until the blue color disappears, calculate the normality of the stock solution.

Normality of thiosulphate solution

The normally of this solution N, is calculated as follows

$N = Mx2.80/V$

Where

V = volume in ml of thiosulphate used and

M = mass in g of potassium iodate.

Sodium thiosulphate titrant —0.01N

Dilute 100ml of stock thiosulphate solution to 1 litre with freshly boiled distilled water.

Standardized sulphite solution for preparation of working sulphite –TCM solution

Dissolve 0.30 g of sodium metabisulphite ($NaHSO_3$) or 0.40g of sodium sulphite (Na_2SO_3) in 500 ml of freshly boiled and cooled distilled water. Sulphite solution is unstable: it is therefore important to use water of the highest purity to minimize this instability. This solution contains the equivalent of 320-400 ug/ml of SO_2. The actual concentration of the solution Is determined by adding excess iodine and back titrating with standard sodium thiosulphate solution. To back titrate pipette out 50ml of the 0.01N iodine solution into each of the two 500ml iodine flasks A (blank) and B (sample). To flask A (blank) add 25 ml of distilled water and into B (sample) measure 25ml of sulphite solution. Stopper the flasks and allow to react for 5 min. By means of a burette containing standardized 0.01N thiosulphate titrate each flask in turn to a pale yellow color. Then add 5 ml of starch solution and continue the titration until the blue color disappears.

Working sulphite —TCM solution

Measure 2ml of the standard sulphite solution into 100 ml volumetric flask by pipette and bring to mark with 0.04 M TCM. Calculate the concentration of sulphur dioxide in the working solution in ug of sulphur dioxide per millilitre. This solution is stable for 30 days if kept in the refrigerator at 5°C otherwise, prepare fresh, when needed.

Calculation

$$C = (V1 - V2) \times N \times 32000 \times 0.02/25$$

Where

C = concentration of SO_2 solution in ug/ml in ml;

V1 = volume in ml of the thiosulphate used for blank

V2 = volume in ml of the thiosulphate used for sample

N = normality of thiosulphate used for sample;

3200 = milli equivalent weight SO_2 ug;

25 = volume of standard sulphite solution ml;and

0.02 = dilution factor

Purified Para Rosaniline stock solution –0.2 percent concentration

The Para Rosaniline dye shall have a wavelength of maximum absorbance at 540 nm when assayed in a buffered solution of 0.1 M sodium acetate –acetic acid; the absorbance of the reagent blank which is temperature – sensitive to the extent of 0.015 absorbance unit /°C shall not exceed 0.170 absorbance unit at 22°C with a 1 cm optical path length, when the blank is prepared according to the prescribed analytical procedure and to the specified concentration of the dye; the calibration curve shall have a slope of 0.03+0.002 absorbance unit /ug SO_2 at this path length when the dye is pure and the sulphite solution is properly standardized.

Preparation of Stock solution

Dissolve 0.5g of Para Rosaniline chloride in 100 ml distilled water. Keep it for 2 days and filter the solution the filtrate solution is stable for 3 months if stored in the refrigerator.

Working para rosanline solution

Add 15ml concentrated hydrochloric acid to 10ml stock Para Rosaniline solution and dilute to 250ml with distilled water in a 250ml volumetric flask. It may be stored at room temperature in an amber coloured bottle for one to two weeks, if stored in a refrigerator.

Sample collection

The air is drawn through 10ml of absorbing solution (0.IM sodium tetrachloromercurate) at the flow rate of 0.5 l/m for 8 hours. The sample is carried to laboratory for analysis.

Determination

For each set of determinations prepare a reagent blank by adding 10ml of unexposed TCM solution to a 25ml of volumetric flask. Prepare a control solution by measuring 2ml of working sulphite -TCM solution into a 25ml volumetric flask by pipette. To each flask containing sample or control solution or reagent blank add 1ml of 0.6percent Sulphamic acid and allow to react for 10 min. to destroy the nitrite resulting from oxides of nitrogen. Add 2ml of 0.2 percent formaldehyde solution and 5ml of Para Rosaniline solution. Start a laboratory timer that has been set for 30 min. Bring all flask to volume with freshly boiled and cooled distilled water and mix thoroughly. Within 30 to 60 minutes determine the absorbance's of the sample reagent blank and the control solution at 560mm using cells with a 1cm path length.

Use distilled water (not the reagent blank) as the optical reference. This is Important because of the color sensitivity of the reagent blank to temperature changes, which may be induced in the cell compartment of a spectrophotometer. Do not allow the colored solution to stand in the absorbance cells, because a film of dye may be deposited. Clean cells with the help of alcohol and pipe-cleaner after use. If the temperature of the determinations does not differ by more than 2^0C from the calibration temperature, the reagent blank should be within 0.03 absorbance unit of the y- intercept of the calibration curve. If the reagent blank differs by more than 0.03 absorbance unit from that found in the calibration curve, prepare a new curve.

Calibration curve

Measure by graduated pipette amounts of the working TCM solution (such as 0, 0.5, 1,2,3 and 4ml) in to a series of 25ml volumetric flasks. Add sufficient TCM solution to each flask to bring the volume to approximately 10ml. Then add the remaining ragents as described above (determination). Plot the absorbance against the total concentration in microgram sulphur dioxide for the corresponding solution. The total micrograms sulphur dioxide in the solution equals the concentration of the standard in micrograms sulphur dioxide per millimeter times the millimeter of sulphite solution added (µg SO_2 = µg ml SO_3 per ml x ml added). A linear relationship should be obtained, and the y-intercept should be within 0.03 absorbance unit of the zero standard absorbance. Determine the slope of the line of best fit calculate its reciprocal and denote as B, the calibration factor. This factor can be used for calculating results provided there are no radical changes in temperature or pH. At least one control sample containing known concentration is recommended to ensure the reliability of this factor.

Conversion of Volume

$$Vn = Vx\ (P/760)\ x\ (298/t+273)$$

Where

Vn = volume of air at 25°C and 760 mm Hg, 1

V = volume of air sampled, 1

P = barometric pressure, mm Hg

T = temperature of air sampled °C

When sulphite solutions are used to prepare calibration curves, compute the concentration of SO_2 (C) in micrograms per cubic meter, in the sample as follows

$$C = [(A-A_o) \times 10^3 \times B / Vr] \times D$$

Where

A = sample absorbance

A_o = reagent blank absorbance

10^3 = conversion of litres to cubic meters;

Vr = the sample corrected to 25 °C and 760 mm Hg

B = Calibration factor; ug/ absorbance unit

D = dilution factor; for 30 min and 1 hr samples D= 1; for 24 hr samples D= 10

SO_2, ppm = $\mu g\ SO_2/m^3 \times 3.82 \times 10^{-4}$

Estimation of Oxides of Nitrogen (NO_x) [IS: 5182 Part VI, 1992 (Jacob and Hochheiser)]

Requirements

Reagents

(a) Absorbing solution of NO_x

It is prepared by 4.0g of NaOH and 1g of sodium arsenite in 1000 ml of distilled water.

(b) Sulphanilamide solution

20g of sulphanilamide dissolved in 700 ml of distilled water and 50 ml of concentrated phosphoric acid (85%) and then dilute the solution to 1000ml.

(c) Hydrogenperoxide

0.gm of H_2O_2 (30%) is diluted to 250ml of distilled water.

(d) 1-naphthyl ethyl diamine hydrochloride (NEDA)

0.5 g of NEDA is dissolved in the 50ml of distilled water.

(All the reagents were stored in the amber bottle at refrigerated temperature for a week).

(e) Standard nitrite solution

The amount of sodium nitrite is calculated by following ways:

$$G = \frac{1.5 \times 100}{A}$$

where G amount in g of sod. nitrite
A Assay percent of sodium nitrite used
1.5 Gravimetric factor in converting NO_2 into sodium nitrite.

Sodium nitrite (assay 97%) 1.5464 g is dissolved in the 1000 ml of distilled water. Stock solution contains. 1000 ìg NO_2/ml.

Procedure

(a) Calibration curve

2.5 ml of stock sodium solution (1000 µg/ml) is diluted to 100 ml with distilled water. The solution contained 25µg NO_2/ml. The standard curve is prepared by taking different concentration of NOx ranges 0.5-5.0 µg NO_2/ml. The standard solution is further processed as per procedure described below for unknown sample.

(b) Sample collection

The air is drawn through 10 ml of absorbing solution (0.4% NaOH solution) at the flow rate of 0.5 l/m for 8 hours. The sample is carried to lab for analysis.

(c) Analysis

10ml of sulphanilamide is added into 10.0 ml of sample followed by the addition of 1.0 ml of H_2O_2 and 1.4 ml of NEDA solution with thorough mixing after addition of each reagent. Blank is prepared in the similar fashion. Absorbance is recorded at 540mm on the spectrophotometer after 30 minutes.

Calculation

$$NOx\ (\mu g/m^3) = \frac{(\mu g\ NO_2/\ 10ml)}{0.82 \times V_S}$$

Where

V_S	=	Volume of air sampled in liters
0.82	=	overall average efficiency
		Concentration of nitrogen dioxide is calculated as ppm nitrogen dioxide
NO_2 (ppm)	=	($\mu g\ NO_2/m^3$) x 5.32 x 10^{-4}

Polycyclic Aromatic Hydrocarbons (PAHs)

Collection of sample: Suspended Particulate Matter (SPM) sample are collected from pre identified locations using high volume samplers operated at a rate of 1.5m^3 /min. The SPM concentration are determined by collecting the particulate matter for 24hr on pre-weighed grassfire filter of 20x25 cm size and reweighed after sampling in order to determine the mass concentration of particles collected. The concentration of particulate matter in ambient air is then computed on the net mass content divided by the volume of sample. SPM is collected and brought to the laboratory in black polythene to avoid photo degradation of PAHs.

Extraction: Filter paper is carefully placed into soxhlet apparatus, already wrapped with black carbon paper to avoid photo degradation during extraction procedure (EPA,1996). PAHs are extracted with 300ml dichloromethane for 16 hours .the extract is dried under reduced pressure. The dried extract is dissolved in the 5 ml of cyclohexane.

Clean up: 10mm inner diameter of chromatography column is prepared by 10g of activated silica gel in the methylene chloride. The slurry is placed into column on the glass wool and allowed to settle down. A uniform bed is prepared. The anhydrous sodium sulphate is placed to avoid moisture from air. Column is pre-eluted with 40ml of pentane to remove any contamination in column materials and the eluted solvent is discarded. The cyclohexane containing extract is poured into column. The column is run by 25ml of solvents mixture (dichloromethane and pentane 2:3). The elute is dried under reduced pressure and dissolved in cyclohexane and the volume is made up to 10ml in volumetric flask with the same solvent.

Analysis: The cleaned extract is applied on HPLC under following analytical conditions to have a good reproducible chromatogram for analysis of naphthalene anthracene ,acenepthylene, fluorine, fluoranthene, benzo(ghi)pyrene, acenepthylene, pyrene, chrysene, benzol(k)fluorene, benzo(a)pyrene and phenanthrene .

Instrument: HPLC shimadzu make model LC 10D C^{18} (250 x 4mm) column length (E, merck made); Detector UV/Visible (SPD-10) Mobile phase: Acetonitrile and water (HPLCgrade) in the ratio of 70:30 Flow rate :1.75ml/min.

Calculation

Concentration of PAHs (ppm) = [Area of unknown x conc. of PAH (ng) x Vol. of sample made up] / [Area of known x Vol. of injected sample (ul) x wt. of sample (gm.)]

Actual amount of PAHs (ng/m^3) = [Conc. of PAHs (ng/mg) x wt. of sample (mg)]/ Vol. of air (m^3)

Formaldehyde *(HCHO)[NIOSH, 1997]*

Reagents

Absorbing solution

Double distilled water

Chromotropic acid

Chromotropic acid 0.1g(4,5-dihydroxy-2, 7-naptha-lenedisulfonic acid disodium salt) is dissolved in 10.0ml of distilled water

Iodine (0.1 N approximately)

Potassium iodide 25.0g is dissolved in 25ml of distilled water and 12.7g of iodine is added to this finally diluted to 1000ml with distilled water.

Iodine (0.01N)

100ml of stock iodine solution is diluted to 1000ml with distilled water.

Standardization of iodine solution

0.01N iodine solution is standardized by 0.01 N sodium thiosolution.

Starch

Paste of 1g soluble starch is prepared by adding 2.0ml of distilled water. The paste is added to 100ml of boiling water. 5ml of cooled chloroform is added.

Sodium carbonate buffer solution

80g of anhydrous sodium carbamate is dissolved in 500ml of distilled water. Then 20 ml of glacial acetic acid is added slowly and made up to 1000 ml with distilled water.

Sodium bisulphite (1%)

1g of sodium bisulphite solution is dissolved in 100ml of distilled water.

Formaldehyde Solution A (1 mg/ml)

Formaline solution (37%) 2.7ml is diluted to 1000 ml distilled water. The solution is standardized as given below:

Formaldehyde standard solution B (10 μg/ml)

1.0ml of standard A is diluted to 100ml with distilled water

Standardization of formaldehyde solution

10.0ml of (1%) of sodium bisulphite and 1ml of starch solution is added in the iodine flask containing 1 ml of formaldehyde solution and in blank solution having 1.0mg distilled water. Then the solution is titrated by 0.1 N of Iodine to a dark blue colour. The excess of iodine is destroyed by 0.05N of sodium thiosulphate. Then solution is titrated by 0.01N Iodine solution. At the end of titration faint blue color appears. The excess of inorganic bisulphite is completely oxidized to sulfate and the solution is ready for the assay of formaldehyde bisulphite addition product.

The 25.0 ml of chilled sodium carbonate buffer is added in a chilled flask containing the product, which finally titrated with 0.01 N iodine solution.

0.01N Iodine = 0.15mg formaldehyde

Procedure

Standard curve

Standard curve is prepared by taking different concentration of formaldehyde solution B (10 μg/ml) ranging from 0.5 to 5.0 μg/ml and diluted by 4.0ml of distilled water. Standard solution is further processed as per procedure described below for unknown sample.

Sampling

Two impingers were attached in the series containing each 20.0ml of distilled water. The air is drawn for 1 hour at flow rate of 1 lit per minute sample is carried to the lab for analysis.

Analysis

The 4.0ml of aliquot is taken from each sampling solution into separate test tubes and the blank tube contained 4.0ml of distilled water, to it was added 1.0ml of Chromotropic acid. 6.0ml of conc. H_2SO_4 is cautiously added very slowly and allowed to cool at room temperature. The absorbance is recorded on the spectrophotometer at 580mm.

Calculation

V = F x T

Where

F= average flow rate of gas

T = Time

Vs = V x P x 298

760 (T+273)

Where V = volume of air in litre during air sampling period

P = Atmospheric pressure in litre during air sampling period

T = Temperature recorded in ^{0}C during air sampling period

$C_T = C_A \times F_A + F_B \times C_B$

Where C_T = Total ug of formaldehyde in the sample

C_A and C_B = Formaldehyde concentration in ug of sample aliquot taken in impinger A and B from calibration curve

F_A and F_B = Respective aliquot factor = sampling solution volume in ml. / ml. of aliquot used

Concentration of Formaldehyde at 25^{0}C and 760 mm Hg ppm (vol.) = C_t x 24.47

Vs.M.W

Where Vs = liter of air samples at standard conditions

M.W. = Molecular weight of formaldehyde = 30.03

Plant Sample Analysis

Pigments

1. Preparation of Homogenate

For pigment analysis 0.5gm of fresh leaves are crushed in pre chilled mortar and pestle with 10 ml of 80% cold acetone and centrifuged to 5000 rpm for 15 minutes

2. Estimation of Pigments

Chlorophyll (Arnon 1949, Strain *et. al.* 1971), Phaeophytin (Vernon et.al. 1960), and Carotenoids (Duxbury *et. al.* 1956)

The supernatant (First) obtained after centrifugation is decanted into a 10ml volumetric flask covered with black carbon to avoid chlorophyll degradation and its volume made up to 10ml with 80% acetone. The residue is extracted with 60% acetone and again centrifuged at 2000 rpm for 10 minutes. The supernatant (second) is decanted into a 10ml volumetric flask and it's volume made up to 10ml with 60% acetone. Both supernatants (first and second) are then mixed in a 25 ml volumetric flask and the total volume made up to 25 ml with

80% acetone. The optical density of the extract is measured at 480, 510, 645, 655, 663 and 666 nm wavelengths on a spectrophoto-meter. The contents of chlorophyll a and b, phaeophytin and carotenoids are calculated and expressed in mg/gm fresh weight of leaves using the following formulae:

Chl a (mg/gm fresh weight) = $12.3\ D_{663} - 0.86\ D_{645} / d \times 1000 \times w$

Chl b (mg/gm fresh weight) = $19.3\ D_{645} - 0.86\ D_{663} / d \times 1000 \times w$

Total Chlorophyll (mg/gm fresh weight) = Chl a + Chl b

Phaeophytin (mg/gm fresh weight) = $6.75\ D_{666} + 26.03\ D_{655} / d \times 1000 \times w$

Carotenoids (mg/gm fresh weight) = $7.6\ D_{480} - 1.49\ D_{510} / d \times 1000 \times w$

Where

V = volume of extract in ml

d = length of light path in cm

w = weight of fresh leaves in gm

D = optical density at different wavelengths

3. Estimation Of Protein (Lowry et. al.,1951)

Protein content in the leaves of the plants is estimated by colorimetric method as described by Lowry *et. al.* in 1951.

Reagents

Following reagents are prepared for the analysis:

Reagent A - 2% Na_2CO_3 in 0.1 N NaOH

Reagent B - 0.5% $CuSO_4.5H_2O$ in 1% sodium potassium tartarate

Reagent C (Alkaline cooper solution) - 50 ml of regent A is mixed with 1 ml of reagent B

Reagent D - 1part Folin phenol + 2 part Double distilled water

Preparation of test solution

100mg fresh leaves are homogenized in 10ml 10% cold trichloroacetic acid solution and centrifuged at 10,000 rpm for 10 minutes. Supernatant is decanted and the pellets are dissolved in 1 N NaOH by warming on water bath for 10 minutes and again centrifuged at 5,000 rpm for 10 minutes. Supernatant is collected and to 0.5ml of clear supernatant, 5ml of reagent C is added and mixed thoroughly. Solution is kept for 10 minutes and then 0.5ml of reagent D is added. Solution is again kept for 10 minute. A blue color appears and its optical density is recorded at 600nm on a spectrophotometer. The protein content (mg/ gm fresh weight) is measured from the standard curve prepared from the Bovine albumin protein.

Preparation of standard BSA solution

The standard solution for protein estimation is made with the help of protein bovine serum albumin (BSA). For this, 1 mg/ml of BSA is prepared by dissolving 100mg of BSA in distilled water. Then 1-2 drops of 0.1N NaOH are added to it. Different volumes of BSA – 0.1, 0.2, 0.3, 0.4, and 0.5ml, are taken in different test tubes and their volume made upto 2ml with distilled water. Alkaline copper reagent is added to each. After half an hour 0.5ml of Folin's reagent is added to them and their absorbance's read at 660mm. Later a standard graph is prepared by plotting these values against different concentration values.

Peroxidase

Preparation of crude extract

10 % homogenate of fresh plant leaves is prepared by crushing 1 gm of leaf sample in 10ml of phosphate buffer with the help of pre chilled pestle and mortar. Homogenate is centrifuged at 3000 rpm for 15 minutes in cold and the supernatant is used as enzyme source.

Preparation of Reaction mixture

Reaction mixture volume of 3ml is taken which contains 0.1 M potassium phosphate buffer at pH 7, 5 mM guaiacol and 2 mM H_2O_2.

Analysis

Reaction is started by adding 0.01ml of homogenate to the 3ml of reaction mixture and the increase of absorbance at 470mm is measured continuously for 3 minutes on UV-visible Cintra 20 spectrometer. The rate of peroxidation of guaiacol to tetraguaiacol is calculated using a molar extinction coefficient 26,600 $M^{-1}cm^{-1}$. Activity is calculated as fresh weight as international enzyme unit and the one enzyme unit (U) is defined as the amount of enzyme catalyzing the oxidation of 1 µmol guaiacol min^{-1} (Srivastava *et. al.*, 1972, Angelini and Fedrico, 1989, Puccinelli *et. al.*, 1998).

Heavy Metals Analysis

Glassware's

The corning glassware's be used and initially washed with tap water and should be immersed in a bath containing nitric acid for 48 hours. After soaking in nitric acid, these glassware's should rinsed several times with double distilled water and finally with deionized water and air-dried.

Reagents

The working metals standard solution be prepared on the day of analysis by suitable dilution of stock solution of metals to be analysed (1000 ppm of Sigma Chemicals, USA) with deionized water.

Air

Collection of sample: Suspended particulate matter (SPM) samples be collected from different locations using HVS and RDS operated at a rate of 1.1-1.3 m^3/min. The SPM concentrations are determined by collecting the particulate matter for 24 hours on pre-weighed EPM 2000 filter of 20 x 25 cm size. Filter paper is reweighed after sampling in order to determine the mass concentration of lead particles collected (Janssens & Dams, 1973). The concentration of metal in ambient air is computed in µg/m^3 by dividing the net mass content with the volume of air (Gajhate and Hassan, 1999).

Sample Preparation

Three circles of 30 mm radius be punched from each filter paper and be digested with HNO_3 in conical flask. The solution should then be filtered into another conical flask through Whatman no. 42 filter paper and the insoluble residue on the filter paper be rinsed with 10ml of 10% nitric acid. A blank be prepared in a similar fashion. The filtered digested solutions were evaporated to almost dryness. The residue should be dissolved in 1% of HNO_3 and made up in 25ml volumetric flask with 1% HNO_3. Finally Atomic Absorption Spectrophotometer (AAS) should be used for analysing digested samples. Three replicates be taken and the average values be used for calculating the results.

Plant

Leaf samples of Ashok, Peepal, Neem, Gulmahar, Amaltas and Mango trees should be collected from locations and finely chopped and oven dried at 65 °C for 24 hour. 0.5 gm of dry leaf material be pulverized and digested in a mixture of nitric and perchloric acid. Samples after clear digestion, be filtered with Whatman 42 and volume be made to 25ml by adding 1N HNO_3 and analyzed by AAS.

Soil

Soil sample should be collected from surface soil (15 to 20cm depth.) at each location. Soil samples are air-dried, sieved, ground and mixed thoroughly and kept in an oven at 65 °C till constant weight (24 hours). 1 g of the oven dried soil sample be digested in a 4 : 1 mixture of nitric and perchloric acid. Samples after digestion are filtered with Whatman 42 and finally the volume be made to 25ml with 1N HNO_3 and analyzed by AAS.

Analysis

All the three type of samples be analyzed on AAS and the sensitivity of AAS should be as follows:

Optimum working range of detection (OWRD) in µg/ml is 0.06-0.15; Cu be estimated at l_{3247}, slit 0.5mm, OWRD is 0.03-10 µg/ml; Zn be estimated at l_{2139}, silt 1mm, OWRD is 0.1-2.0 µg/ml; Ni be estimated at l_{232}, silt 0.2 mm, OWRD is 0.1-2.0 µg/ml; Cr be estimated at $l_{357.9}$, silt 0.2 mm, OWRD is 0.06-5 µg/ml, Pb be estimated at l_{217}, silt 1 mm, OWRD is 0.01-3.0 µg/ml and Cd be estimated at $l_{228.6}$, silt 0.5 mm, OWRD is 0.02-3 µg/ml.

Calculation

Conc. of metal in collected sample (μg/gm) =

$$\frac{\text{Conc. of metal (μg/ml) x Vol. of digested sample (ml)}}{\text{Wt. of sample used for digestion (gm)}}$$

Conc. of metal in ambient air (μg/m^3) =

$$\frac{\text{Conc. of metal in sample collected}}{\text{Vol. of air sampled (m}^3\text{)}}$$

Chapter - 13

STANDARDS TO CONTROL AIR POLLUTION

Pollution is continuously increasing in India. To control ambient air pollution CPCB has prescribed National Ambient Air Quality Standards

National Ambient Air Quality Standards

Pollutants	Time-Weighted Average	Concentration in Ambient Air			Method of Measurement
		Industrial Areas	Residential Rural & Other Areas	Sensitive Areas	
Sulphur Dioxide (SO_2)	Annual Average*	80 µg/m³	60 µg/m³	15 µg/m³	- Improved West and - gaek eUltraviolet Fluorescence
	24 hours**	120 µg/m³	80 µg/m³	30 µg/m³	
Oxides of Nitrogen as (Nox)	Annual Average*	80 µg/m³	60 µg/m³	15 µg/m³	- Jacob & Hochheiser Modified (Na-Arsenite) Method
	24 hours**	120 µg/m³	80 µg/m³	30 µg/m³	- Gas Phase Chemiluminescence
Suspended Particulate Matter	Annual Average*	360 µg/m³	140 µg/m³	70 µg/m³	- High Volume Sampling, (SPM) (Average flow rate not less than 1.1 m3/ minute).
	24 hours**	500 µg/m³	200 µg/m³	100 µg/m³	
Respirable Particulate Matter (RPM) (size less than 10 microns)	Annual Average*	120 µg/m³	60 µg/m³	50 µg/m³	- Respirable particulate matter sampler
	24 hours**	150 µg/m³	100 µg/m³	75 µg/m³	

Pollutants	Time-Weighted Average	Concentration in Ambient Air Industrial Areas	Residential Rural & Other Areas	Sensitive Areas	Method of Measurement
Lead (Pb)	Annual Average*	1.0 μg/m³	0.75 μg/m³	0.50 μg/m³	· ASS Method after sampling using EPM 2000 or equivalent Filter paper
	24 hours**	1.5 μg/m³	1.00 μg/m³	0.75 μg/m³	
Ammonia	Annual Average*	0.1 mg/ m³	0.1 mg/ m³	0.1 mg/m³	
	24 hours**	0.4 mg/ m³	0.4 mg/m³	0.4 mg/m³	
CarbonMonoxide (CO)	8 hours**	5.0 mg/m³	2.0 mg/m³	1.0 mg/ m³	· Non Dispersive Infra Red (NDIR) Spectroscopy
	1 hour	10.0 mg/m³	4.0 mg/m³	2.0 mg/m³	

* Annual Arithmetic mean of minimum 104 measurements in a year taken twice a week 24 hourly at uniform interval.

** 24 hourly/8 hourly values should be met 98% of the time in a year. However, 2% of the time, it may exceed but not on two consecutive days.

Emission Standard

The Parameters Determining Emission From Vehicles

» Vehicular Technology

» Fuel Quality

» Inspection & Maintenance of In-Use Vehicles

» Road and Traffic Management

The significant role of vehicles in urban air pollution cannot be denied. The need to reduce vehicular pollution has led to emission control through regulations together with increasingly eco-friendly technologies. In India it was only in 1991 that the first stage emission norms came into force for petrol vehicles and in 1992 for diesel vehicles. From April 1995 mandatory fitment of catalytic converters in new petrol passenger cars sold in the four metros of Delhi, Kolkata, Mumbai and Chennai along with supply of Unleaded Petrol (ULP) was affected. Availability of ULP was further extended to 42 major cities and now it is available throughout the country.

The emission reduction achieved from pre-89 levels is over 85% for petrol driven and 61% for diesel vehicles from 1991 levels. In the year 2000 passenger cars and commercial vehicles will be meeting Euro I equivalent India 2000 norms, while two wheelers will be meeting one of the tightest emission norms in the world. Euro II equivalent Bharat Stage II norms are in force from 2001 in 4 metros of Delhi, Mumbai, Chennai and Kolkata. Since India embarked on a formal emission control regime only in 1991, there is a gap in comparison with technologies available in the USA or Europe. Currently, we are behind Euro norms by few years, however, a beginning has been made, and emission norms are being aligned with Euro Standards and vehicular technology is being accordingly upgraded. Vehicle manufactures are also working towards bridging the gap between Euro Standards and Indian Emission Norms (Siamindia .com).

International Emission Standard for Vehicles

Emission fall into three categories

1. European Emission system (Euro System).
2. Japanese Emission system.
3. American Emission system.

The European Emission Standards are followed by most of the countries, including India.

EURO-I are those that should have been applicable from year 2000.

EURO-II- are those that should have been applicable from year 2005.

Pollutant	**Euro – I Petrol**	**Diesel**	**Euro - Ii Petrol**	**Diesel**
C O g/km	2.72	2.72	2.2	1.00
H C + Nox g/km	0.97	0.97 (IDI)	0.57	0.70
1.36 (DI)	——	——	——	——
P M g/km	——	0.14 (IDI)	——	——
0.19 (DI)	——	——	——	——

The Indian Pollution Standard For Vehicles

The Central Motor Vehicles Rules-115(II) 1989 says that the pollution standards laid down for carbon monoxide and inspected in petrol vehicle should be as under :

» 2 & 3 wheelers-Idling Carbon monoxide emission by volume: 4.5%

» 4 wheelers-Idling Carbon monoxide emission by volume : 3.0%

For Diesel vehicles the emission standards for inspection of smoke density are as under:

Maximum Smoke	**Density Light absorption co-efficient (l/m)**	**Bosch Units**	**Hartidge Units**
(i) Full load at 60 to 70% of maximum engine rated rpm declared by the manufacturer.a	3.25	5.2	75
(ii) of Free acceleration	2.45	—	65

The vehicle emission standards of India are given in Table A and B below:

Table A: Emission standards for gasoline light duty vehicles (in gm/km)

Year	**CO**	**HC**	**HC+NO_x**
1991 (a)	14.3-27.1	2.0-2.9	-
1996 (a, b)	8.68-12.40	-	3.0-4.36
2000 (b)	2.72	-	0.97

(a) Depends upon engine capacity; (b) No crankcase emission; Evaporation emission 2.0 gm/test max.

Table B: Emission standards for heavy diesel vehicles (in gm/kw.hr)

Year	**CO**	**HC**	**NO_x**	**PM**
1992	14	3.5	18	-
1996	11.2	2.4	14.4	-
2000	4.5	1.1	8	0.36

(*Source:* Transport Fuel Quality for Year 2005, page 69, Central Pollution Control Board, New Delhi)

According to the Society of Indian Automobile Manufacturers Emission Standards for petrol and diesel driven vehicles are:

Petrol Vehicles				
Three – Wheelers (g/km)				
Year	**CO**	**HC**	**HC+Nox**	
1991	30-Dec	12-Aug	-	-
1996	6.75	-	5.4	-
2000	4	-	2	-
2005(BS II)	2.25	-	2	(DF =1.2)
Two – Wheelers (g/km)				
Year	**CO**	**HC**	**HC+Nox**	
1991	30-Dec	12-Aug	-	-
1996	4.5	-	3.6	-
2000	2	-	2	-
2005(BS II)	1.5	-	1.5	(DF =1.2)
Car (g/km)				
Year	**CO**	**HC**	**Nox**	**HC+Nox**
1991	14.3 - 27.1	2.0-2.9		
1996	8.68 - 12.4			3.00 - 4.36
1998*	4.34 - 6.20			1.50 - 2.18
2000	2.78			0.97
B.S II	2.2			0.5
B.S II	2.2 - 5.0			0.5 - 0.7
B.S III	2.3	0.2	0.15	
B.S III	2.3 - 5.22	0.20 - 0.29	0.15 - 0.21	

Upto 6 seaters (A) and GVW upto 2.5 tonnes More than 6 seaters (B) and GVW upto 3.5 tonnes (A) (B)

* For Catalytic Converter Fitted vehicles

Diseal Vehicles Diseal Vehicles (GVM Upto 3.5 Tonnes) (g/km) Engine Dynamometer						
Year	**CO**	**HC**	**Nox**	**HC+Nox**	**PM**	
1992	14	3.5	18			
1996	11.2	2.4	14.4			
2000	4.5	1.1	8		0.36/ 0.61 #	
B.S II	4	1.1	7		0.15	For Four Wheelers only
Or (g/km) Chasis Dynamometer						
Year	**CO**	**HC**	**Nox**	**HC+Nox**	**PM**	
1992	17.3 -32.6	2.7 -3.7				Light Duty Vehicles
1996	5.0 -9.0			2.0-4.0		
2000	2.72 -6.90			0.97-1.70	0.14 -0.25	
B.S II	1.0 -1.5			0.7-1.2	0.08 -0.17	For Four Wheelers only
B.S II (2005)	1.00			0.85	0.10	For 2 & 3 Wheelers, Appropriate DF
B.S III	0.64 -0.95		0.50 -0.78	0.56 -0.86	0.05 -0.10	
CARS (g/km) Chasis Dynamometer						
BS II	1.0			0.7	0.8	(A)
B.S.II	1.0 -1.5			0.7 -1.2	0.8 -0.17	(B)
B.S.III	0.64		0.50	0.56	0.05	(A)
B.S.III	0.64 -0.95		0.50 -0.78	0.56 -0.86	0.05 0.10	(B)

Diseal vehicles (GVM > 3.5 Tonnes) (g/kwh)						
Year	**CO**	**HC**	**Nox**	**HC+Nox**	**PM$**	**Smoke (m-1) $**
1992	14.0	3.5	18			
1996	11.20	2.40	14.4			
2000	4.5	1.1	8.0		0.36 0.36	
BS.II	4.0	1.1	7.0		0.15	
B.S.III	2.1	0.7	5.0		0.10 -0.13	0.8

For Engines with Power exceeding 85 kw/ For engines with power upto 85 kw
* For engines having swept volume < 0.75 I per cylinder & rated power speed > 3000 rpm
$ For Diesel Vehicles Only

BS III Norms w.e.f. 1st April '05 in 11 Cities

For CNG Vehicles HC to be replaced by NMHC: NMHC = HC x (1-K/100)
K - % Methane Content in NG
For LPG Vehicles HC to be replaced by RHC: RHC = 0.5 x HC

Source : www.siamindia.com

References

Abollino Ornella, Aceto Maurizio, Malandrino Mery, Mentas Edoardo, Sarzanini Corrado Sarzanini And Barberis Renzo, 2002, Distribution And Mobility of Metals In Contaminated Sites. Chemometric Investigation of Pollutant Profiles, *Environ, Pollut.,* 119: 17-193.

Addison P.A., Malhotra S.S. And Khan A.A., 1984, Effect of Sulphur Di Oxide On Woody Boreal Forest Species Grown On Native Soils And Tailings., *J. Environ. Qual.* 13:333-336.

Agarwal M., Singh S.K., Singh J., And Rao DN., 1991, Bio-monitoring of Air Pollution Around Urban And Industrial Sites, *J. of Env. Bio.* 211.

Agarwal S., Tiwari S.L., 1996, Effect of Industrial Air Pollution On Leaf Area And Dry Weight Ratio And Photosynthetic Pigments of Some Tree Species, *Environment And Ecology* 14(4): 818-820.

Agency for Toxic Substance and Disease Registry (ASTDR), 1988, The nature and extent of lead poisoning in children in the US., A Report to Congress, US Department of Health and Human Services.

Agrawal Madhoolika, Najma Khanam (Dept Bot, Banaras Hindu Univ, Varanasi 221005), 1997, Variations In Concentrations of Particulate Matter Around A Cement Factory, *Indian J Environ Hlth* 39(2), 97- 102.

Akhter M.S. And Madany I.M, Heavy Metals In Street And House Dust In Bahrin, 1993, *Water Air And Soil Pollut.* 66: 111-119.

Aksoy A. And Ozturk M.A., 1997, Nerium Oleander L. As A Biomonitor of Lead And Other Heavy Metal Pollution In Mediterranean Environments, *Sci. Total Environ.* 205: 145-150.

Albasel N. and Cottenie A., 1985, Heavy metal concentration near majoe highways, industrial and urban area in Belgian Grassland, *Water Air Soil Pollution,* 24: 103-109.

Alder J.M. and Carey P.M., 1989, Air toxics emissions and health risks from mobile sources, 82 [nd] annual meeting of the air and waste association, Anaheim, California.

Alfani A., Maisto G., Iovieno P., Rutigliano F.A., Bartoli G., 1996, Leaf Contamination By Atmospheric Pollutants As Assessed By Elemental Analysis of Leaf Tissue, Leaf Surface Deposit And Soil, *Journal of Plant Physiology* 148: 243:248.

Ali E.A., Nasralla M.M. And Shakour A.A., 1986, Spatial And Seasonal Variation of Lead In Cairo Atmosphere, *Environmental Pollution*, 11(B): 205-210.

Allerd E.N. *et. al.*, 1989, Short term effects of CO exposure on the exercise performance of subjects with coronary artery disease, *New England Journal of Medicine,* 321: 426-432.

Amundson R.G., Walker R.B., Legge A.H., 1986, Sulphur Gas Emission In The Boreal Forest: The West White Court Case Study. Vii. Pine Tree Physiology. *Water Air Soil Pollution* 29: 129-147.

Anbazhagan M., Krishnamurthy R., And Bhagwat KA., 1989, Environmental Pollution, 58: 125.

Angelini Riccardo And Rodolfo Fedrico, 1989, Histochemical Evidences of Polyamine Oxidation And Generation of Hydrogen Per Oxide In The Cell Wall, *J. Plant Physiol.,* 135: 212-217.

Anthropogenic Emissions In India, Ohio Super Computer, Columbus, *www.Osc.Edu/Research/Pcrm/Emissions/Fuels.html*

Antoniadis V., Alloway B.J., 2002, The Role of Dissolved Organic Carbon In The Mobility of Cd, Ni And Zn In Sewage Amended Soils, *Environmental Pollution,* 117(3): 515-521.

Archer V.E., 1990, Air pollution and fatal lung disease in three Utah Countries, *Arch. Environ. Hlth.,* 45:325-334.

Arnon D.I., 1949, Copper Enzymes in isolated chloroplasts polyphenoloxidase in beta vulgaris, *Plant Physiol.,* 24: 1-15.

Arnow W.S. *et. al.*, 1981, Aggravation of angina pectoris by two percent carboxy hemoglobin, *Am. Heart.* J., 101:154-157.

Asada K.and Kiso K., 1973, Initiation of aerobic oxidation of sulphite by illuminated spinach chloroplast, *Eur. J. Biochem.,* 33: 253-257.

Avol E.L. *et. al.*, 1987, Short term respiratory effects of photochemical oxidant exposure in exercising children, *J. Air. Pollut. Cont. Assoc.,* 37:158-162.

Ash S. *et. al.*, Maternal weight gain, smoking and other factors in pregnancy as predictors of infant birth weight in Sydney Women, *Australian and New Zealand Journal of Obs. and Gynae.*, 29:212-219.

Bag S.C., Bhattacharya S.K., Das B.K., Saha S.C., Saiyed H.N. (Natl Inst Res Jute Allied Fibre Techno, Indian Coun Agricl Res, 12, Regent Park, Calcutta 700040), 1997, Characterization of Dust Collected Frons Jute Mill Environment, *India J Indl Med,* 43(1) 8-10.

Bagley S.T., Baumgard K.J., Gratz L.G., Jonson J.H., Leddy D.G., 1996, Characterization of fuel and after treatment device effects and diesel emissions, *Health Effects Institute (HEI), Research Repot* No. 76.

Balsberg Pahlsson A.M., 1989, Effect of Heavy Metal And So_2 Pollution On The Concentration of Carbohydrates And Nitrogen In Tree Leaves, *Canadian Journal of Botany,* 62: 2106-2113.

Baltrop D., Children and environmental lead, Hepple P (eds) Lead in the environment: proceedings of a conference, London, UK, Institute of Petroleum, 52-60.

Barman S.C. And Lal M.M., 1994, Accumulation of Heavy Metals (Zn, Cu, Cd, And Pb) In Soil And Cultivated Vegetables And Weeds Grown In Industrially Polluted Fields, *J. Environ. Biol.* 15(2): 107-115.

Bauer L.I. de, Hernandez T.T. and Manning W.J, 1985, *JAPAC*, 35: 838.

Beckett K.P., Freer-Smith P.H. And Taylor G., 1998, Urban Woodlands: Their Role In Reducing The Effects of Particulate Pollution, *Environ. Poll.,* 99: 347-360.

Beg M.A.A., 1990a, Report On Status of Air Pollution In Karachi: Past Present And Future, Pakistan Council of Scientific And Industrial Research, Karachi.

Beg M.U., Mohd. Farooq, Bhargava S.K., Kidwai M.M., Lal M.M., 1990b, Performance of Trees Around A Thermal Power Station, *Environment And Ecology*, 8(3): 791-797.

Bhadoria A.K.S., 1994, Heavy Metals In Soil, Their Problems And Remedies, In V.S. Tomar (Ed) Water And Nutrient Management In Soils, Proc. Summer Institute Held At Dept. Soil Sci. Agri. Chem. J.N.K.V.V., Jabalpur May 30 to June 18, 218-221.

Bhaduria Seema, 1999, Biokinetics of In-Vitro Oil Biodegradation In Soil, *J. Env. Polln.,* 6(2&3): 133-140.

Bharti N. And Singh R.P., 1993, Growth And Nitrate Reduction By *Sesamum Indicum Cv* Pb-I Respond Differently to Lead, *Phytochemistry*, 33: 531-535.

Bhattacharjee S. and Mukherjee A.K., 1994, Influence Cd and Pb on physiological and biochemical responses of Vigna uterguiculata (L) on cell injury, piment, sugar, nucleic acid and paeroxidase content, *Pollution Res.,* 13(3): 279-286.

Bhattacharya, S. And Mitra A.P., May 1998, (Eds.) Greenhouse Gas Emissions In India For The Base Year 1990: *Global Change Scientific Report No. 11,* Published By The Center For Global Change, National Physical Laboratory, New Delhi- 110 012.

Biswajit R. Bhaumik G.C. (Dept Chem, Regl Engg. Coll, Durgapur 713209, West Bengal), 2001, Some Gaseous Pollutants In Durgapur, An Industrial City of India. *J Env Polln,* 8(4) (2001), 383-384.

Blackwell, 1992, Urban Air Pollution In Mega Cities of The World: Published On Behalf of The World Health Organization And The United Nations Environment Program.

Boa W., Zhu Z., 1997, *Agro-Environmental Protection*, 16 (In Chinese).

Bobrov R.A.., 1955, The Leaf Structure of Poa Annua With Observations On Its Smog Sensitivity In Los Angels Country, *American J. Bot.,* 4: 467-474.

Bonneau M,, Fink S,, Rennenberg H,, 1995, Methods to Assess the Effect of Chemicals On Ecosystem, *Scope,* 53.

Boubel R.W., Fox D.L., Turner D.B., Stern A.C., 1994, Fundamentals of Air Pollution, Academic Press, San Deigo.

Bowen H.J.M., Environmental Chemistry of the Elements, Academic Press, 1979, New York.

Bragaloni M., Garrou E., Puccinelli P., 1993, Verifica De Una Metodica Monitoraggio Dell'inquinamento Atmosferico Con Piante Di Interesse Forestale In Ambiente Urbano, *Biologica Ambientale-Bollettino Cisba.* 31: 9-15.

Bragaloni M., Puccinelli P., Garrou E., 1995, Monitoraggio Dell'inquinamento Atmosferico Nell'ambiente Urbano Suburbano Di Torino Mediante Valutazione Delle Perossidasi Fogliari Di Pinate Sempreverdi. Biologi. Italini *(Organo Ufficale Dell'ordine Nazionale Dei Biologi)* 7(XXV): 39-46.

Braun E. And Lewark S., 1986, Jahrringanalysen In Bohrkernen Aus Den Berliner Forsten . *Holz Als Roh- Und Wertstoff,* 44: 326-327.

Brown D.M., Stone V., Findlay P., Macnee W., Donaldson K., 2000, Increased Inflammation And Intracellular Calcium Caused By Ultra Fine Carbon Black Is Independent of Transition Metals Or Other Soluble Components, Occup. *Environ. Med.,* 57(10): 685-691.

Brun L.A., Maillet J., Hinsinger P., Pin A., 2001, Evaluation of Cu Availability To Plants In Cooper Contaminated Vineyard Soils, *Environmental Pollution* 111(2): 293-302.

Buma T.J., 1960, Bull. Am. Meterol. Soc., 41:357-360.

Cadle R.D., 1966, Particles In The Atmosphere And Space, Reinhold Publishing Corporation, New York.

Calabrese E.J. *et.al.,* 1981, A Review of human health effects associated with exposure to diesel fuel exhaust, *Environ. Int.*, 5:473-477.

Canada, Federal-Provincial Advisory Committee On Air Quality. 1987. Review of National Ambient Air Quality Objectives For SO_2: Desirable And Acceptable Level. Ottawa: Environment Canada.

Carlson T. and Bazaz B., 1977, Growth reduction in Americal Seymore (*Planatanus occidentalis*) caused by Pb-Cd interaction, *Environ. Pollut.,* 12: 243-253.

Carreras H.A., Pignata M.L., 2001, Comparison Among Air Pollutants, Meteorological Conditions And Some Chemical Parameters In The Transplanted *Lichen Usnea Amblyoclada*, *Environmental Pollution* 111(1): 45-52.

Caselles J., Colliga C. and Zornoza P., 2002,Evaluation of Trace Element Pollution From Vehicle Emissions In Petunia Plants. *Water Air And Soil Pollution,* 136: 1-9.

Ceburnis, D., And E. Steinnes. Conifer Needles As Biomonitors of Atmospheric Heavy Metal Deposition: Comparison With Mosses And Precipitation, Role of The Canopy." *Atmospheric Environment*, 24(25):4256-4271.

Central Pollution Control Board, 1999, Ambient Air Quality Program, 1998-1999.

Central Pollution Control Board, 2000, Transport Fuel Quality For Year 2005, pp. 210,, New Delhi.

CETSEB 1989, Relatorio De Qualidade Do Ar No Estado De Sao Paulo 1988 Serie Realtorios, Companhia De Tecnologia De Saneamento Ambiental, Sao Paulo, Brazil.

CETSEB, 1990, Relatorio De Qualidade Do Ar No Estado De Sao Paulo 1989 Serie Relatorios, Companhia De Technologia De Saneamento Ambiental Sau Paulo. Brazil.

Chaaban F.B., Ayoub G.M. And Oulabi M., 2000, A Preliminary Evaluation of Selected Transport Related Pollutants In The Ambient Atmosphere of The City of Beirut, Lebanon, *Water Air And Soil Pollution*, 126: 53-62.

Chamberlian A.C. And Monteiths, J.L. (Eds.) 1975, The Movement of Particles In Plant Communities In Monteith J.L. (Eds), Vegetation And The Atmosphere Vol. 1, Academic Press London, 155-203.

Chan T.L., Dong G., Cheung C.S., Leung C.W., Wong C.P. And Hung W.T., 2002, Monte Carlo Simulation of Nitrogen Oxides Dispersion From A Vehicular Exhaust Plume And Its Sensitivity Studies. *Atmospheric Environment* 35(35): 6117-6127.

Chang L., Muller F.J., Ultman J. 1991, Alveolar epithelial cell injuries by subcutaneous exposure to low concentration of ozone correlate with cumulative exposure, toxicol. *Appl. Pharmacol.*, 109:219-234.

Chelani Asha B., Padke K.M., Gajghate D.G., Hasan M.Z. (National Environ Engg. Res Inst, Nagpur 440020), 2002, Status of PM_{10} In Four Coastal Cities of India. *Proc Natl Conf Polln Prev Contl India* : 2-3 March 2002, Nagpur, 59-63.

Chen Hong And Cutright Teresa, 2001, Edta And Hedta Effects On Cd, Cr And Ni Uptake By Helianthus Annus, *Chemosphere* 45: 21-28.

Choie D.D. and Richter G.W. 1980, Effects of lead on kidney, lead toxicity Singhal R.L. and Thomas J.A. (eds), Urban and schwarzenberg, Baltimore, 337-350.

Chow J.C., Watson J.G., Lowenthal D.H., 1996, Sources And Chemistry of Pm10 Aerosol In Santa Barbara County, Ca. *Atmospheric Environment* 30: 1489-1499.

Citizens Fifth Report 1999. Anil Agrawal & Sunita Narain (Eds.), Published By Centre For Science And Environment, 41, Tughlakabad Institutional Area, New Delhi-110 062.

Clarie L.C., Adrianod.C., Sajwas K.S., Abel S.L., Thoma D.P., Driver J.T., 1991, Effects of Selected Trace Metals On Germinating Seeds of Six Plant Species. *Water Air Soil Pollution*, 59:231-240.

Clarkson Dt. *et.al.,* 1993 : Membrane and Long Distance Transfer of Sulphati; Sulphur Nutrition and Assimilation in Higher Plants, Academic Publ. The Haque, Netherlands.

Clavert J.G., Su F., Bottenhem J.W. and Strausz O.P., *Atmos. Environ.*, 12:197-226.

Clean Air Environmental Governance-4, Igidr, Mumbai, 1999.

Colombo J.C, Skorupka C.N, Bilos C. And Rodriguez Presa M.J., 1999, Sources, Distribution And Variability of Air Borne Trace Metals In La Plata City Area, Argentina. *Environmental Pollution*, 104: 305-314.

Cool M., Marcoux F., Paulin A. And Mehra M.C., 1980, Metallic contamionants in street soils of Moncton, New Brunswick, Canada, *Bull. Environ. Contam. Toxicol.*, 25: 409-415.

CPCB, 1995, Ambient Air Quality Status And Statistics: 1993-1994, *National Air Quality Monitoring* (NAQM)/2/1995-96.

Cracker L.E.., 1972, Influence of O_3 On RNA And Protein Content of Lemna Minor L., *Environ. Pollut.,* 3: 319-323.

Critteden P.D.., And Read D.J., 1979, The Effect of Air Pollution On Plant Growth With Specila Reference To SO_2 Iii Growth Studies With Lolium Multiforum Linn. And Dactylis Glomerata L., *New Phytol*, 83: 645-651.

Croxford B., Penn A., Hillier B., 1996, Spatial Distribution of Urban Pollution: Civilizing Urban Traffic, *Science of Total Environment*, 190: 3-9.

Crump D., 1996, Indoor Air Quality- The Alspac Study, *Proceedings of 63rd National Society For Clean Air Environmental Protection Conference And Exhibition*, National Society For Clean Air, Brighton.

CVGMO, 1986, Review of atmospheric pollution and harmful substance emissions in Moscow region for 1985, Moscow Center for Hydrometeorology and Environmental Observations, Moscow.

Darral N.M.., 1986, The Sensitivity of Natural Photosynthesis In Several Species To Short Term Fumigation With Sulphue Di Oxide., *J. Exp. Bot.* 37: 1313-1322.

Davis M.J. and Svendsgard D.J., 1987, Lead and child development, *Nature*, 329:297-300.

Davis R.D. and Beckett P.H.T., 1978, Upper critical levels of toxic elements in plant II. Critical levels of Cu in young barley, wheat, rape, lettuce and rye grass and of Ni and Zn in young barley and rye grass, *New Phytol*, 80: 23-32.

Del Maestro R.F., 1980, An Approach to free radicals in medicine and biology, *Acta Physiol Scand.*, 492: 153-168.

Dennis A.N.N., Fraser Mathew, Anderson Stephson And Allen David, 2002, Air Pollutant Emissions Associated With Forest, Grassland and Agricultural Burning In Texas, *Atmospheric Environment* 36(23): 3779-3792.

Demji K.S. and Richters A., 1989, Reduction of T lymphocyte subpopulations following acute exposure to 4 ppm NO_2, *Environ. Res.,* 49:217-224.

Department of Environment, 1995a, Expert Panel On Air Quality Standard: Particles, HMSO, and London.

Dietrich K.N. *et. al.,* 1990, Neurobehavioral effects of fetal lead exposure: the first year of life, 1987b in Smith M., Grant L.D., Sors A. (eds), Lead exposure and child development: an international assessment, Lancaster, UK, MTP Press.

DDF, 1989, Programa integral contra la contaminaction atmosferica programa de emergenica, proyectos, Departmento del distrito federal Mexico DF.

Di Giulio, R.T. and C.J. Richarson, 1987, Effects of atmospheric deposition on red spruce: A free radical based approach, *Annual report to spruce Bioomall*, Penn.

Dockery D.W., Pope A., Xu X., Spengler J.D., Ware J.H., Fay M.E., Ferris B.G. and Speizer F.E., 1993, An association between air pollution and mortality in six US cities, *New England Journal of Medicine*, 329: 1753-1759.

Dockery D.W. *et. al.,* 1989, Effects of inhalable particles on respiratory health of children, *Am. Rev. Respir. Dis.,* 139:587-594.

Dollard G.J., 1986, Glasshouse experiments on the uptake of foliar applied lead, *Environ. Pollut.*, 40(A): 109-119.

Donalson K., Stone V., Clouter A., Renwick L., Macnee W., 2001, Ultra Fine Particles, *Occup. Environ. Med.*, 58(3): 211-216.

Dudka S. And Miller W.P., 1999, Accumulation of Potentially Toxic Elements In Plants And Their Transfer To Human Food Chain, *J. Environ. Sci. Health* B34(4): 681-708.

Dugger W.M. Jr., Taylor O.C.., Cardiff E. And Thomoson C.R., 1962, Stomatal Action In Plants As Related To Photochemical Oxidants, *Plant Physiol.*, 37: 487-491.

Duxbury A.C. and Yentsch C.S., 1956, Plankton pigment monograph, *J. Mar. Res.,* 15: 19-101.

Eisler R., 1993, Zinc Hazards To Fish, Wildlife And Invertebrates: A Synoptic Review, United States Fish And Wild Life Service, *Biological Report* 10: 1-106.

Emberson L.D., Ashmore M.R., Cambridge H.M., Simpson D. And Tuovinen J.P., 2000, *Environ Poll.,* 109: 403.

Environment Australia, 2001, State of Knowledge Report: Air Toxics And Indoor Air Quality In Australia, Environment Australia 2001.

EPA, 1998, The Annual Assessment Report of The Air Pollution Control In Taiwan Area For 1997 (In Chinese).

European Environment Protection Agency (EEPA), 1998, Europe's Environment The Second Assessment, Elsevier, Oxford.

Faiz A., Sinha K., Walsh M. And Varma A. (Eds), 1990, Automotive Air Pollution Issues And Options For Developing Countries, World Bank Policy And Research Working Paper 492, World Bank, Washington DC.

Faiz. Iqbal., Christopher.S. Weaver. And Walsh.P. Michel,1996, (Eds.) Air Pollution From Motor Vehicles.

Farmer A.M., 1993, The Effect of Dust On Vegetation – A Review, *Environment Pollution* 79: 63-75.

Farmer A.M., 1995, Reducing The Impact of Air Pollution On The Natural Environment, Joint Nature Conservation Committee, Peterborough.

Fatoki O.S., Ayodelee.T., 1991, Zinc And Cooper Levels In Tree Barks As Indicators of Environmental Pollution, *Environmental International* 13: 369-373.

Fenn M.E., Byterowicz A., 1997, Summer Through Fall And Winter Deposition In The San Bernardinao Mountains In Southern California, *Atmospheric Environment*, 31: 293-310.

Fergusson J.E. And Kim N.D., 1991, Trace Elements In Streets And House Dust: Sources and Speciation, *Sci., Total Envn.,* 100: 125-150.

Firket M., 1936, Fog along Meuse valley, *Trans. Faraday. Soc.*, 32:1192-1197.

Frank M.B., 1991, Control of motor vehicle emissions, the US Experience Critical Reviews in *Environmental Control*, 21(5,6):333-340.

Freedman B., 1995, Environment Ecology (2nd Edition), New York, London Academic Press.

Freer-Smith P.H., Broadmeadow M.S.J., 1996, Urban Woodland And The Benefits For Local Air Quality, Aboriculture Advisory And Information Service Research Note, Farnham.

Freguson C., Kasamas H., 1999, Risk Assessment For Contaminated Sites In Europe Vol 2, Policy Framework, Lqm Press Ed, Nottingham.

Friedland A.J., 1990, The Movement of Metals Through Soil An Ecosystem, In Heavy Metal Tolerance In Plants: Evolutionary Aspects (Ed), A.J. Shaw CRC Press, Boca Raton, Florida 7-19.

Fuel Cell Bus Development In India, October 1999 - September 2000 Undp/Gef Pdf-B Project, Min. of Non-Conventional Energy Sources (Govt. of India), Bharat Heavy Electricals Limited.

Fulton M. *et. al.*, 1987, Influence of blood lead on the ability and attainment of children in Edinburgh, *Lancet.*, 1(8544):1221-1226.

Gajhate D.G., Hasan M.Z., 1999, Ambient Lead Levels In Urban Areas, *Environmental Contamination And Toxicology*, 62: 403-408.

Gardea- Torresdey J.L., Polette L., Arteaga S., Tiemann K.J., Bibb J., Gonzalez J.H., 1996, Proceedings of the 11th annual EPA Conf. On hazardous Waste research, Edited L.R. Erickson, D.L. Tillison, S.C. Grant, J.P. McDonald, Albuquerque, NM, 660.

Garg K.K. And Varshney C.K., 1980, Effect of Air Pollution In The Leaf Epidermis At The Submicroscopic Level, *Experentia*, 36: 1364-1366.

Garshick E. et. al., 1987, A Case study of lung cancer and diesel exhaust exposure in railroad workers, *Am. Rev. Respir. Dis.,* 135:1242-1248.

Ghauri B.M.K., Salam M., Mirza M.I., 1988, A Report On Assessment of Air Pollution In The Metropolitan Karachi, Space Science Division, Pakistan Space Upper Atmosphere Research Commission (Suparco), Karachi.

Gobierno De La Republica Mexico, 1990, Programa Integral Contra La Contamicion Atmosferics. Gobierno De La Republica Mexico DF.

Goldsmith J.R. and Friberg L.T., 1976, Effect of air pollution on human health in Air pollution 3rd ed. Vol. II, the effects of air pollution, Academic Press, New York, 457-610.

Government of Japan, 1990, Japanese Performance of Energy Conservation And Air Pollution Control, Environment Agency, Government of Japan, Tokyo.

Goyal S.K, Gupta H.K., Lohitesh M.D., Chalapti Rao C.V. (Natl Environ Engg. Res Inst, Nagpur 440020), 2002, Effect of Seasonal Variations On Dispersion of Air Pollutants – A Case Study. *Proc Natl Conf Polln Prev Contl India* : 2-3 March 2002, Nagpur, 65-69.

Grant L.D. and Davis J.M., 1990, Effects of low level lead exposure on pediatric neurobehavioral and physical development: current findings and future directions in *Smith M and Graant L.D. and sors A., (eds.),* 13ead exposure and child development: An international Assessment Lancaster, UK, MTP press.

Greenhalghl. And Worpole K., 1995, Park Life: Urban Parks And Social Renewal, Comedia And Demos, Gloucester And London.

Greenhouse Gas Emissions in India for the base year 1990: Global Change Scientific report No. 11, May 1998, *Bhattacharya, S. and Mitra A.P. (eds.)* Published by the Centre for Global Change, National Physical Laboratory, New Delhi- 110 012.

Green Spaces Task Force Response To Interim Report, 2001, A Joint Response On The 21 Jan 2001, Submission From: Aboriculture Association, Institute of Chartered Foresters, National Aborists Association, National Urban Forestry Unit, North West Tree Officer Group, The Woodland Trust.

Gregory P.H., 1973, The Microbiology of The Atmosphere, Clarke Doble And Brendon, Plymouth.

Grunhagel L., Klein H. And Jager Hj., 1981, Langzeiteinwirkungen Von Frieilandrelevanten SO_2 Und Cadmiumkonzentrationen An Pflanzen. Nachweis Und Wirkung Forstshadlicher Luftverunreinigungen, *Xi Internatiolen Arbeitstagung Forstlicher Rauchss chadenssach verstadiger,* 1-6 September 1980, Graz, Austria, 309-317.

Guderian R., Klumpp G. And Klumpp A., 1998, Effect of SO_2, O_3 And No_x Singly And In Combination On Forest Species In *Ozturk M. (Ed.),* Plant And Pollutants In Developed And Developing Countries. Ege Univ. Press Izmir, Turkey 235-242.

Gupta V.K. And Dixit M.L., 1992, Influence of Soil Applied Cadmium On Growth And Nutrient Composition And Plant Species, *J. Indian Soc. Soil Sc.,* 40: 878-880.

Gutpa V.K. And Patolia B.S., 1990, Zinc-Cadmium Interaction In Wheat, *Indian Soc. Soil Sc.,* 38(3): 452-457.

Hall D.E., King D.J., Morgan T.D.B., Baverstock S.J., Heinze P., Simpson B.J., 1998, A Review of recent literature investigating the measurements of automotive particulate: the relationship with environmental aerosol, air quality and health effects, Society of Automotive Engineers, Technical Paper No. 982602.

Hall. P., 1994, The Innovative City, OECD Conference, Melbourne 1994.

Hampton C.V., Pierson W.R., Harvey T.M., Updegrove W.S. and Marano R.S., 1982, Hydrocarbon gases emitted from vehicles on the road, a qualitative gas chromatography/ mass spectrometry survey, *Environ Sci. Technol.*, 16:287-298.

Hampton C.V., Pierson W.R., Scchuetzle D., and Harvey T.M., 1983, Hydrocarbon gases emitted from vehicles on the road, detemination of emission rates from diesel and spark ignition vehicles, *Environ Sci. Technol.*, 17:699-708.

Harrison R.M., Jones M., Collinis G., 1999, Measurements of the physical properties od particles in the urban atmosphere, *Atmospheric Environment,* 33: 309-321.

Harrison R.M., Laxen D.P.H., and Wilson S.J., 1981, Chemical associations of lead, Cadmium, Copper and Zinc in street dusts and roadside soils, *Environ. Sci. Tehnol.*, 15: 285-297.

Harvey P.G *et. al.,* 1984, Blood lead, behavioral and intelligence test performance in preschool children, *Sci. Total Environ.*, 40:45-60.

Heagle A.S.., Miller J.E., Sherrill D.E., 1994, A White Clover System To Estimate Effects of Tropospheric Ozone On Plants, *J. Environ. Qual.,* 23: 613-621.

Heath Effect Institute (HEI), 1988, Air pollution, the automobiles and public health, National Academic Press Washington DC, 19.

Heath R.l., 1988, Biochemical Mechanisms of Pollutant Stress In Heck WW., Taylor Oc. And Tingley D.T. (Eds.), Assessment of Crop Loss From Air Pollutanta, Elsevier New York, 259-286.

Heath Robert L., 1998, Alteration of chlorophyll in plant upon air pollution exposure in National Academy Press, Biological markers of air pollution stress, 347-356.

Hebel J.R. *et. al.,* 1988, Dose response of birth weight to various measures of maternal smoking during pregnancy, *J. Clin. Epidemiol.,* 41:483-489.

Helbwachs G., 1983, Effect of Air Pollution On Vegetation In: W. Holzner, Mja Werger And I Ikusimaa (Eds.) Mans Impact On Vegeation DrW. Junk Publishers, London, 55-67.

Herscbach C., De Kok L.J., Rennerberg H., 1995a, Net Uptake of Sulphate And Its Transport To The Shoots In Spinach Plants Fumigated With $H2_S$ and SO_2 : Does Atmospheric Sulphur Affect The Inter Organ Regulation of Sulphur Nutrition?, *Botanica Acta*, 108: 41-46.

Herscbach C., De Kok L.J., Rennerberg H., 1995b, Net Uptake of Sulphate And Its Transport To The Shoots In Tobaco Plants Fumigated With $H2_S$ And SO_2, *Plant And Soil*, 175: 75-84.

Heylin M., *Chem. Eng. News,* Feb 11, 1985.

Heywood John B And Sher Eran, 1999, The Two-Stroke Cycle Engine, Sae International Publication.

Holzoworth G.C., 1974, Climatological aspects of the composition and pollution of atmosphere, *WHO Tech. Notes139*, WHO Geneva.

Ho Y.B. and Tai K.M., 1988, Elevated levels of lead and other metals in roadside soil and grass and their use to monitor aerial metal depositions in Hong Kong, *Environ. Pollut*, 49: 37-51.

Horrocks R.W., 1994, Light-Duty Diesels - An Update On The Emissions Challenge' Proc. Inst Mech Engrs, 208: 209, *Journal of Automobile Engg.*, pp 289.

Horvath E., Anderson H., Pierece W. et. al., 1988, Effects of formaldehyde on mucous membranes and lungs, A study of an industrial population, *JAMA*, 259:701-707.

Hovmand M.F., Tjell J.C. and Mosback H., 1983, Plant uptake air borne Cadmium, *Environ. Pollut.*, 30(A): 27-38.

Huffamn E.W.D., Alloway W.H., 1973, Chromium in plants: distribution in tissues organelles and extracts and avalibility of bean leaf chromium to animals, *J. Agric. Chem.*, 21: 982-986.

Hung I.F., Hsin- Fa Fang and Tung-San Lee, 1992, Aliphatic and aromatic hydrocarbons in indoor air, *Bull. Environ. Contam. Toxicol.*, 48:579-584.

Hunter J.G., Vergano O., 1953, Trace element toxicities in oat plants, Ann. Appl. Biol., 40: 761-777.

Hunter T.G. and Vergnano O., 1953, Trace element toxicities in Oats, *Ann. App. Biol.*, 40: 761-777.

Imai M. *et. al.*, 1985, A survey of health studies of photochemical air pollution in Japan, *J. Air Poll. Cont. Assoc.*, 35: 103-108.

Imiwari Swarnlata, Bansal Samidha, Rai Shashi, 1992, Responses of Ficus Relegiosa Exposed To NO_2, *Indian J Env Toxico*, 2(2), 21-26.

Indian Standard: Motor Gasoline - Specification (Third Revision) Ics 43.060.01;75.160.20, Bureau of Indian Standards, Manak Bhawan, 9, Bahadur Shah Zafar Marg, New Delhi-110 002.

Interim Report: Winter Smog and Traffic, A govt. Report: *Announcement and index (GRA and I)*, Issue 2, 1996.

International Agency for Cancer Research (IACR), 1982, Benzene and Annex., Some industrial chemicals and dyestuffs, monographs on the evaluation on carcinogenic risk of chemicals to humans, 29.

International Agency for Cancer Research (IACR), 1989, monographs on the evaluation of carcinogenic risk of chemicals to humans, Diesel and gasoline engine exhaust and some nitroarenes, 46.

Ivestigation of Pollutant Profiles, *Environmental Pollution* 119(2): 177-193.

ITRC (Industrial Toxicology Research Center), Repot released on 4th November, 2000, Assessment ambient air quality of Lucknow city during post monsoon season, by Environment monitoring division.

Jain Aruna., 1992, The Status of Chlorophyll Content In Some Herbaceous Plants of Polluted And Non Polluted Sites of Bhopal., *Indian J. Pure Appl. Bio.,* 7(1), 65-70.

Jain P.K. and Surendra M., 1986, Lead in environment from petrol engine vehicles, *Proceedings of national seminar on environmental pollution control and monitoring.*

Janssens N.A.H., Van Mansom D.F.M., Van Der Jagt K., Harseema H., Hock G., 1997, Mass Concentration And Elemental Composition of Air Borne Particulate Matter At Street And Background Locations, *Atmospheric Environment,* 31: 1185-1193.

Jiang L. and Zang D., 1996, *Chongquin Environmental Science* 18: 33 (in Chinese).

Joshi V.H., Rathore S.S. and Arora S.K., 1999, Effect of Chromium on growth and development of cowpea (*Vigna unguiculata*), *Indian J. Environ. Proct.,* 19: 745-749.

Juttner F., 1988, Changes In Monoterpene Concentration In Needles of Polluted And Injured Picea Abies Exhibiting Mottled Yellowing, *Physiol. Plant,* 72: 48-56.

Kabata-Pendias A. And Pendias H., 1984, Trace Elements In Soil And Plants, CRC Press, Boca Raton, Fl.

Kabata-Pendias A. And Pendias H., 1994, Trace Elements In Soil And Plants, 2nd Ed., CRC Press, Boca Raton, Fl.

Kafka Zdenek And Kuras Mecislav, 1997, Heavy Metals In Soil Contaminated From Different Sources In Ecological Issues And Environmental Impact Assessment, Advances In Environmental Control Technology Series, Paul N. Cheremisinoff, Gulf Publishing Company 175-205.

Karjalainen R., Jokinen J., Laine E. And Martikanen V., 1992, Changes In Enzyme Activity As An Indicator of Air Pollution Stress In Woody Plants, In Roy S., Karamlampi L. And Hanninen O. (Eds.) 7th International Bioindicators Symposium And Workshop On Environmental Health, Abstract Book 35, Kuopio University Publ. C., *Natural And Environmental Science 7,* Finland.

Katiyar Vinita and Dubey P.S., 2001, Sulphue dioxide sensitivity on two stage of leaf development in a few tropical tree species, *Indian J. Environ. Toxicol.,* 11(2): 78-81.

Katoh T., Kasuya M., Kagamimork S., Kazuka S. And Kawano S., 1989, Effects For Air Pollutants On Tannin Biosynthesis And Predation Damage In Cryptomeria Japonica, *Phytochem.,* 28: 439-445.

Katyal J.C., Das S.K. and Sharma K.L., 1992, Interaction of Zn in soil and their management, *Fert. News* Vol. 37(4): 27-33.

Keller T., 1974, The Use of Peroxidase Activity For The Use of Monitoring And Mapping Air Pollution Areas, *Eur. J. For Path.* 4, 11-19.

Keller T., 1980, The Effect of Continuous Springtime Fumigation With So_2 And Co2 Uptake And Structure of Annual Rings In Spruce, *Can J, For. Res.,* 10: 1-6.

Keller T., Schwager H., 1977, Air Pollution and Ascorbic Acid, *Eur. J. Forest Pathol.* 7(6): 338-350.

Kellogg W.W., Cadle R.D., Allen E.R., Lazarus A.I., Martell E.A., 1972, *The Sulphur Cycle Science* 175: 587-596.

Kelly B.I., 1986, Eskom Specialist Environ. *Investigations. Report* Number Trr/ N86/101: 41.

Khan A.M., Pandey V., Shukla J., Singh N. Yunus M., Singh S.N. And Ahmad K.J., 1990, Effect of Thermal Power Plant Emission On *Catharanthus Roseus L., Bull. Environ. Contam. Toxicol.,* 44: 865-870.

Khan H.V., 1981, Concentration of heavy metal in Bangkok metropolitan area, *J. Environ. Sci. Hlth.,* 16(A): 637.

Khandekar R.N., Kelkar D.N., Vohra K.G., 1980, Lead, Cadmium, Zinc, copper and Iron in the atmosphere of greater Bomaby, *Atmos. Environ.,* 14: 457-461.

Khare P., Kulshrestha U.C., Saxena A., Kumar N., Kumari K.M. And Srivatava S.S., 1996, The Source Apportionment of Particulate Matter Using Enrichment Factor And Principal Analysis, *Indian J. Environ. Hlth.,* 38(2)86-94.

Kittelson D.B., 1998, Engines and nonparticles: a review, *J. Aerosol Science,* 29: 575-588.

Kleiner K., 1997, Clean Air Plans Steeped In Confusion, *New Scientist,* 153(2069) 8.

Kleinman M.T. *et. al.,* 1990, Health effects of acid aerosols formed in atmospheric mixtures, *Environ. Hlth Perpect.,* 79:137-145.

Klump A., Klump G., And Domingos M., 1996, *Gefarstiffe- Reinhaltung Der Luft,* 56:27.

Kohert R.J., Amundson R.G.Lawrence J.A., 1986, Evaluation of Growth And Yield of Soyabean Exposed To Ozone In The Field, *Env. Poll. Series A,* 41:219-234.

Komeiji T., Akoki K., Koyama I., Okitat.,1990. Trends of Air Quality And Atmospheric Deposition In Tokyo. *Atmos. Environ.* 24a: 2099-2103.

Kostner B., Cyzan F.C And Lange Ol., 1990, An Analysis of Needle Yellowing In Healthy And Chlorotic Norway Spruce (*Picea Abies*) In Forest Decline Area In The Fitchtelgebirge (*N.E. Bavaria*), *Trees*, 4: 55-67.

Kozhauharov S.I., Petrova A.V., Veroukova L., 1985, Some Aspects of Using Plants As Indicators of Environmental Pollution. *In Symposium On Biomonitoring State of Environment,* Indian National Science Academy, New Delhi-110002, 98-105.

Koziol M.J. And Whaltey F.R., 1984, Gaseous Air Pollutants And Plant Metabolism, Butterworth London :480.

Krishnayappa N.S.R. And Bedi S.J., 1986, Effect of Automobile Lead Pollution On *Cassia Tora* And *Cassia Occidentalis* L., *Environ. Pollut.* 40: 221-225, 1986.

Krupnick A., Sebastian I., 1990 Issues In Urban Air Pollution: Review of Beijing Case. *Environment Working Paper No. 31*, World Bank Washington DC.

Kulshrestha K., Farooqui A., Srivastava K., Singh K.J., Ahmad K.J., Behl H.M., 1994, Effect of Diesel Exhaust Pollution On Cuticular And Epidermal Features of Lantana Camara L. And Syzgium Cuminii L. (Skeels), *Journal of Environmental Science And Health,* 29(A): 301-308.

Kumar G, Singh R.P. And Sushila, 1993, Nitrate Assimilation And Biomass Production In Sesamum Indicum L. Seedlings In A Lead Enriched Environment, *Water Air And Soil Pollut.* 66: 163-171.

Kumar Naresh And Jayshree, 1999, Effect of SO_2 Pollution On The Metabolism of Polyalthia Longifolia, *J. Nature Conserv.* 11(2): 241-244.

Laborans G.F., Novillo A., 1996, Toxicity And Bioaccumulation of Cadmium In Olisthodiscus Luteus, *Water Resources* 30(1): 57-62.

Lagerweff J.V. and Specht A.W., 1970, Contamination of roadside soil and vegetation with Cadmium, Nickel, Lead and Zinc, *Environ. Sc. Technol.,* 4: 583-586.

Landolt W., Guecheva M. and Bucher J.B., 1989, The Spatial distribution of different elements, in and on the foliage of Norway Spruce growing in Switzerland, *Environ. Pollut.,* 56: 155-167.

Larcher, 1994, Okophysiologie Der Pflanze. Utb, Stuttgart.

Laxen D.P.H., And Thompson M.A., 1987, Sulphur Di Oxide In Greater London. 1931-1985, *Environmental Pollution* 43: 103-114.

Lechowicz M.J., 1987, Resource Allocation By Plants Under Air Pollution Stress: Implications For Plant Pest Pathogen Interactions, *Bot. Rev.,* 53: 281-300.

Lee H., Su Sy, Liu KS and Chou MC, 1994, Correlation between meteorological conditions and mutagenicity of air borne particulate samples in tropical monsoon area from Kcohsiunq city, *Tiwan, Environ. Mot. Mut.,* 23(3):200-207.

Lee J.A., Baxter R. And Emes M.J., 1990, Responses of Sphagnum Species To Atmospheric Nitrogen And Sulphur Deposition, *Bot. J. Linn. Soc.,* 104: 255-265.

Lee T.T., 1965, Sugar Content And Stomatal Width As Related To O_3 Injury In Tobacco Leaves, *Canadian J Bot.,* 43: 677-685.

Lin C.C. and Yang S.H., 1996, *Chinese Journal Agromet*, 3: 183 (in Chinese).

Lin Z.Q., Schuepp P.H., 1995, Trace Metal Concentration In And On Balsam Fir (Abies Balsamea (L) Mill), Foliage In Southern Quebec, Canada, *Water Air and Soil Pollution,* 81: 175-191.

Lippman M, (1989), Effect of ozone on respiratory function and structure, *Ann. Rev. Public Hlth.,* 10:49-67.

Lohr V.I., Pearson Mims, C.H., 1996, Particulate Matter Accumulation On Horizontal Surfaces In Interiors: Influence of Foliage Plants, *Atmospheric Environment,* 30: 2565-2568.

Lonneman W.A., Seila R.L. and Butalini J.J., 1978, *Environ. Sci. Technol.*, 12: 459-463.

Lowry O.H., Rosebourgh N.J., Farr A.L. And Randall R.J., 1951, Protein Measurements With Folin Phenol Reagent, *J. Biol. Chem.,* 193: 265-275.

LSS, 1990a, London Air Pollution Monitoring Network- Fourth Report 1989, Lss, Lwmp/120 London Scientific Services, London.

Lyons T.J., Scott W.D., 1990, Principles of Air Pollution Meteorology Belhaven Press, London.

MacNicol R.G. and Beckett P.H.T., 1985, Critical tissue concentration of potentially toxic elements, *Plant Soil*, 85: 107-130.

Madhumanjari Mandal, Mukherji S., 2000, Changes In Chlorophyll Content, Chlorophyllase Activity, Photosynthetic Co2 Uptake, Sugar And Starch Contents In Five Dicotyledons Plants Exposed To Automobile Exhaust Pollution. *Jounal of Env. Biology* 21(1): 37-41.

Mahaffery K.R. *et. al.*, 1982, National estimates of blood lead levels, US 1972-1980, *New Eng. J. Med.,* 307:573-579.

Malhotra S.S. And Hocking D., 1976, Biochemical And Cytological Effects of SO_2 On Plant Metabolism, *New Phytol,* 76: 227-237.

Malhotra S.S. and Khan A.A., 1984, Biochemical and physiological impact of major pollutants in M. Treshow (ed), Air Pollution and Plant Life, John Wiley, New York, 113.

Mannes F., Altieri A., Boffa A., Bruno F. Federico R., 1987, Early Diagnosis of Injuries In Pinus Pinaster Aiton Treated With Simulated Acid Rain, *Ann. Bot.* 45: 71-79, 1987.

Mannes F., Altieri A., Bruno F. Tripodo P., 1989, Plants As Monitor of Environmental Pollution In Urban System of Rome, In Man And Environment, The Plants Components In Anthropic Systems, Sbi. Working Group On Ecology, Rome 24 Oct.

Mannes F., Federico R., Bruno F., 1986, Peroxidase Activity In Nicotiana Tobacum L, Treated With Simulated Acid Rain, *Phytopath. Medit.* 25: 76-79.

Manning. I., 1980, The Journey To Work. Hornsby, Nsw: George Allen And Unwin Australia, 1978. E. O. Pederson, Transportation In Cities, Network: Program.

Mastral A.M., Callen M.S., 2000, A Review On Polycyclic Aromatic Hydrocarbon Emission From Energy Generation, *Environmental Science And Technology* 34: 3051-3057.

Mathai M., *et. al.,* 1990, Maternal smoking and reduced birth weights, *Australia and New Zealand J. Obs Gynaecol*, 30:33-36.

MCCGNPS, 1991, Year Book of The Air Pollution In Moscow City In Towns And In The Settlements of The Moscow Region For 1990, Moscow Centre For Hydrometerology And Environmental Observations, Moscow.

Mcdonald R.D., Zielinska B., Fujita E.M., Sagebiel J.C., Chow J.C., Watson J.G., 2000, Fine Particles And Gaseous Emissions Rates From Residential Wood Combustion, *Environmental Science And Technology* 34: 2080-2091.

Mcgovern F. And Wilson J., 2000, An Assessment of Global Atmospheric Sulphate Levels, And Levels Measured In Ireland, Dept. of Experimental Physics, University College Dublin, Belfield, Dublin, Environment Institute, Joint Research Centre, Ispra (Via), Italy.

Mcpherson E.G., Nowak D.J., Rowntree R.E., 1994, Chicago's Urban Forest Ecosystem: Results of The Cchicago Urban Forest Climate Project, USDA General Technical Report No-186.

Meadows. D.H., Meadows. D.L., Randers. J., Behrene. W.H., 1972, Limits To Growth, Study By Club of Rome.

Meagher.B., 2000, Phytoremediation of Toxic Elemental And Organic Pollutants, Curr. Opin. *Plant Biol.,* 3: 153-162.

Melhorn H., Gunther S., Schmidt A. and Kunert K.J., 1986, Effect of SO_2 and O_3 on production of antioxidants in conifers, *Plant Physiol.,* 82: 336-338.

Merkert B., 1994, Plant As Biomonitors- Potential Advantages And Problems In D.C. Adraino, Z.S. Chen And S.S. Yang (Eds), Biogeochemistry of Trace Elements, *Science And Technology Letters*, Northwood, NY., 601-613.

Mertz W., 1969, Chromium occurrence and functions in biological systems, *Physio. Rev.,* 49: 163-239.

Mielke H.W., 1991, Lead In Residential Soils, *Water Air And Soil Pollution*, 38: 239-250.

Mishra L.C., 1982, Effect of Environmental Pollution On The Morphology And Leaf Epidermis of Commelina Benghalensis Linn., *Environ. Pollut. (Series A),* 28: 281-284.

Moll G., 1996 Using Geographic Information Systems (GIS) To Analyze The Value of Urban Ecosystems. In Urban Trees – Costing The Benefits, Conference Proceeding, Chartered Institute of Water And Environment Management, Landon.

Monaci F. And Bargagli R., 1997, Barium And Other Trace Metals As Indicators of Vehicle Emission, *Water Air And Soil Pollution*, 110: 89-98.

Monn C.H., Braendii O., Schaeppi G., Schindler C.H., Ackermann Liebrich U., Leuenberger P.H., Sapaldia Team, 1995, Particulate Matter <10μM (Pm10) And Total Suspended Particulates (Tsp) In Urban Rural And Alpine Air In Switzerland, *Atmospheric Environment* 29: 2565-2573.

Morawaska L., 2001, New Directions: Particle Air Pollution Down Under, *Atmospheric Environment*, 35: 1711-1712.

Morawaska L., Bofinger N.D., Kocis L., Nwankwoala A., 1998a, Submicrometer and supermicrometer particles from diesel vehicle emissions, *Environmental Science and Technology*, 32: 2033-2042.

Morawaska L., Thomas S., Bofinger N., Wainwright D., Neale D., 1998b, Comprehensive characterization of aerosols in the sub tropical urban atmosphere: particle size distribution and correlation with gaseous pollutants, *Atmospheric Environment*, 32: 2467-2478.

Morawasks L., Jayaratne E.R., Mengersen K., Jamriska M. And Thomas S., 2002, Differences In Airborne Particle And Gaseous Concentrations In Urban Air Between Weekdays and Weekends, *Atmospheric Environment,* 36(27): 4375-4383.

Morikawa H., Higaki A., Nohno M., Kamada M., Nakata M., Toyohara G. And Fujita K., 1993, Air Pollutant-Philic Plants From Nature In Murata N. (Ed.), *Proceedings of The IXth International Congress On Photosynthesis*, Kluwer Academic Dordrectht, 18: 897-900.

Moroni J.S., Briggs K.G., Taylor G.J., 1991, Chlorophyll Content And Leaf Elongation Rate In Wheat Seedlings A Measure of Manganese Tolerance, *Plant And Soil,* 136: 1-9.

Motor Vehicle Manufacturers of India, 1999, Vehicle population of Delhi, Delhi.

Motto H.L., Danies R.H., Chilko D.M. and Motto C.K., 1970, Lead in soils and plants: its relationship to traffic volume and proximity to highways, *Environ. Sc. Technol.,* 4: 231-237.

Munday P.K.,Timmis R.J., And Walker C.A., 1989, A Despersion Modelling Study of Present Air Quality And Oxides of Nitrogen Concentration In Greater London, Warren Spring Laboratory, Stevenage UK.

Municipal Development Authority (MCD), 1999, State of Delhi, Delhi.

Nagoor S., Vyas A.V., 1999, Physiological And Biochemical Responses of Cereal Seedlings To Graded Levels of Metals. Iii Effects of Copper On Protein Metabolism In Wheat Seedlings, *J. Environ. Bio.* 20(2)125-129.

Naik V.B., Deshpande U.P., 2000, An Evaluation of The Roadside Plants As Bio Indicators of Atmospheric Lead Pollution, *Indian J. Environ. Hlth,* 42(2): 92-93.

Nandi P.K., Agarwal M., Agarwal S.B. And Rao D.N., 1990, Physiological Responses of Vicia Faba Plants To Sulphur Di Oxide, *Ecotoxic. Environ. Safety,* 19: 64-71.

Naomi Tsuku, *et. al.*, 2001, Diesel exhaust effects the regulation of Testicular function in male Fischer 344 rats, *Environ. Hlth.,* 63(A): 115-126.

NAPAP (National Acid Precipitation Assessment Program), 1990, *Effect of Pollution On Vegetation*, Report 18. Washington D.C., Govt. Printing Office.

NAS, 1972, Lead: Ai borne lead in perspective, committee on biological effects of atmospheric pollutants, Washington, D.C.

National Research Council, 1979, Ozone and other photochemical oxidants, National Academy of Sciences, Washington D.C.

National Swedish Institute of Environmental Medicine (NSIEM), 1983, Health risks resulting from exposure to motor vehicle exhaust, *a report to the Swedish Govt., Committee, on automotive air pollution*, NSIEM, Stockholm.

Needhum J.R., Doyal D.M., Faulkner, S.A. And Freeman H.D., 'Technology For 1994, 1989, *SAE Paper* 891949, 1989.

NEERI, 1991a, Air Pollution Aspects of Three Indian Megacities : Delhi, NEERI, Nagpur India.

NEERI, 1991b, Air Pollution Aspects of Three Indian Megacities : Bombay, NEERI, Nagpur India.

NEERI, 1991c, Air Pollution Aspects of Three Indian Megacities : Calcutta, NEERI, Nagpur India.

Newman. Peter, 1997,Urban Ecology, Transportation, Energy And Land Use In Ecological Issues And Environmental Impact Assessment- *Advances In Environmental Control Technology* Series By Paul N Cheremisinoff, 35-96.

Newmann P.W.G. And Kenworthy J.R., 1989, Cities And Automobile Dependence: An International Source Book, Gower, Aldershot.

Niazi G.A., Fleming A.F. and Siziya S., 1989, Blood dyscrasia in unofficial vendors of petrol and heavy oil and motor mechanics in Nigeria, *Trop. Doct.*, 195:55-58.

Ninave S.Y, Chaudhri P.R., Gajghate D.G. And Tarar J.L, 2001, Foliar Biochemical Features of Plants As Indicators of Air Pollution, *Bull. Environ. Contam. Toxicol.*, 67: 133-140.

Norhoa N. and Gaur J.P., 1995, Effects of Cations including heavy metals, on cadmium uptake by Lemna polyrhiza L., *Bio Metals,* 8: 95-98.

Nowak D.J., 1994, Air Pollution Removal By Chicago's Urban Forest In Mcpherson E.G., Nowak D.J., Rowntree R.A., (Eds), Chicago's Urban Forest Ecosystem: Result of Chicago Urban Forest Climate Project, USDA Forest Service, Northeastern Forest Experimental Station, *General Technical Report.*

Nowak, D. J., Mchale, P. J., Ibarra, M., Crane, D.,Stevens. J.C., Luley, C. J., 1997. Modeling The Effect of Urban Vegetation On Air Pollution. In: 22nd NATO/ CCMS International Technical Meeting On Air Pollution Modelling And Its Application, Pp. 276-282. Nato/Ccms, Brussels.

Nriagu J.O., 1996, A History of Global Metal Pollution, *Science*, 272: 223-234.

Nwosu J U, Harding A K, And Linder G, 1995, Cadmium And Lead Uptake By Edible Crops Grown In Silt Loam Soil, *Bull. Environ. Contam. Toxicol.,* 54: 570-578.

NYDEC, 1990, New York State Air Quality Report. Ambient Air Monitoring System *Annual 1989, USEPA, Research*, Triangle Park NC.

Oberdorster G., 2001, Pulmonary Effects of Inhaled Ultra Fine Particles, Arch. Occup. *Environ. Health*, 74(1): 135-151.

Oberdorster G., Ferin J., Gelein R., Ferin J., Weiss B., 1995, Association of particulate air pollution and acute mortality: involvement of ultra fine particles, *Inhalation Technology*, 7: 111-124.

Oberdorster G., Ferin J., Gelein R., Soderholm S.C., Finkelstein J., 1992, Role of alveolar macrophage in lung injury: studies with ultra fine particles, *Environment and Health Prespectives*, 97: 193-199.

Ohio Supercomputer Center, www.Osc.Edu/Research/Pcrm/Emissions/ Partii.html

Okano K., Machida T And Totsuka T., 1989, Differences In The Ability of No_2 Absorption In Various Broad Leaved Tree Species, *Environ. Pollut.,* 58: 1-17.

Oksanen J., Tynnyrinen S. And Kaerenlampi L., 1990, Testing For Increased Abundance of Epiphytic Lichen On The Local Pollution Gradient, *Ann. Bot. Fenn.,* 24: 301-307.

Olbrich K.A., 1990, The Geographic Distribution of Possible Air Pollution Induced Symptoms In The Needles of *Pinus Patula*, CSIR Report 1: 39.

Olesen J.H. *et. al.*, 1984, Occupational formaldehyde exposure and increased nasal cancer risk in man, *Int. J. Can.,* 34:639-644.

Olszyk D.M. And Tingey D.T., 1984, Phytotoxicity of Air Pollutants Evidence For Phyto Detoxification of SO_2 But Not O_3, *Plant Physiol,* 74: 999-1005.

ONEB, 1989, Air And Noise Pollution In Thailand 1989.No. 07-03-33, Office of National Environment Board, Bangkok.

Oren R., Schulze E.D., Werk K.S.., Meyer J., Schneider B.U., Heilmeier H., 1988.Performance of Two Picea Abies (L) Karst Stands At Different Stages of Decline In Carbon Relations And Stand Growth, *Oecologia* (Berl.) 75: 25-37.

Oren R., Werk K.S., Buchmann N And Zimmermann R., 1993, Chlorophyll-Nutrient Relationships Identify Nutritionally Caused Decline In *Picea Abies* (L) Karst Stands, *Can. J. For Res.,* 23: 1187-1195.

Oros D.R., Simoneit B.R.T., 1999, Identification of Molecular Tracers In Organic Aerosols From Temperate Climate Vegetation Subjected To Biomass Burning, *Aerosol Science And Technology* 31: 433-445.

Osonubi O. And Davies W.J., 1980, The Influence of Plant Water Stress On Stomatal Control of Gas Exchange At Different Levels of Atmospheric Humidity, *Oecologia*, 46: 1-6.

Ostro B., 1984, A Search for the threshold in the relation of air pollution to mortality: a reanalysis of data on London mortality, *Environ. Hlth. Perspect.*, 58:397-399.

Pahlsson A.M.B., 1989, Toxicity of heavy metals (Zn, Cu, Cd, Pb) to vascular plants, Lit. Rev., *Water Air and Soil Pollution*, 47: 287-319.

Pais I., Jones Benton J. (Jr.), 1997, The Handbook of Trace Elements, St. Lucie Press, Boca Raton, Florida.

Pandaya K.P. *et. al.*, 1991, An Investigation of environmental impact on health of workers at retail petrol pumps, *Ann. Occup. Hyg.*, 33:347-364.

Pandey D.D., Sinha C.S., Tiwari M.G., 1992, Impact of Coal Dust Pollution On Biomass, Chlorophyll, Nutrient And Grain Characteristic of Wheat., *J Ecobiol.*, 4(1): 19-22.

Panigrahi N.C., Misra B.B., Mohanty B.K., 1992, Effect of SO_2 on chlorophyll content of two crop plants, *Environ. Biol.*, 13(3): 201-205.

Pandey, Pandoy U (Cent Adv Std Bot, Banaras Hindu Univ, Varanasi 221005), 1997, Adaptational Strategy of A Tropical Shrub *Carissa Carandas* L. To Urban Air Pollution. *Environ Monit Assess*, 43(3): 255-265 (Paper Reports The Adaptational Response of A Tropical Shrub Carissa Carandas L. To Urban Air Pollution Stress In Varanasi, India).

Pandit M, Singh A.P., Kapoor J.C. 2001, Measurement of Suspended Particulate Matter And Noise Levels During Blast Trials In A Coal Field, *Pollen Res*, 20(3): 429-433.

Pandya K.P., Rao G.S., Dhasmana A. and Zaidi S.H., 1975, Occupational exposure of petrol pumps workers, *Ann. Occup. Hyg.*, 18:363-364.

Parikh. Jyoti And Kirit Parikh, 1999, Accounting And Valuation of Environment, Volume I, Case Studies From The Escap Region, Un-Escap-Sd/Tecpub/1.

Parkinson G.S., 1971, Benzene in motor gasoline: an investigation into possible health hazards in and around filling stations and in normal transport operations, *Ann. Occup. Hyg.*, 14:145-153.

Pcssaraakli Mohammad, 1999, Handbook of plant and crop stress (2nd edition), Culinary and Hospitality Industry Publications Services.

Pemberton. Max., 2000 Pemberton Associates, Managing The Future- World Vehicle Forecasts And Strategies To 2020, Volume *1: Changing The Patterns of Demand An Automotive World Report.*

Penuelas Joseph And Filella Iolanda, 2002, Metal Pollution In Spanish Terrestrial Ecosystems During The Twentieth Century, *Chemosphere*, 46: 501-505.

Peralta-Videa J.R., Gardea-Torresday, Gomez E., Tieman K.J., Parsons J.G., Carillo G., 2002, Effect of Mixed Cadmium, Cooper, Nickel And Zinc At Different Phs Upon *Alfalfa* Growth And Heavy Metal Uptake, *Environmental Pollution*, 119(3): 291-301.

Peters A., Wichmann H.E., Tuch T., Heinrich J., Heyder J., 1997, Respiratory effects are associated with the number of ultrafine particles, *American Journal of Respiration and Critical Care Medicine*, 155: 1376-1383.

Pfanz H., Dietz K.J., Weinerth J And Opmann B., 1990, Detoxification of So_2 By Apoplastic Peroxidases In Rennerberg H., Brunold C., De Kok Lj. And Stulen I (Ed.), Sulphur Nutrition And Sulphur Assimilation In Higher Plants, Spb, Acad. Publ. The Hauge, 229-233.

Pierre M., Querioz O., 1971, Enzymatic and metabolic changes in bean leavesduring continous pollution by sub necrotic levels of SO_2, *Environ. Pollut.,* 25: 41-51.

Pieser G.D., Yang S.F., 1978, Chlorophyll Destruction In The Presence of Bisulphite And Linoleic Acid Hydroperoxide, *Phytochem.* 17: 79-84.

Plette A.C.C., Hannstra L., Riemsdijk W.H., 1993, Heavy Metal Interactions With Soil Bacteria: In Integrated And Sediment Research: A Basis For Proper Protection, H.J.P. Eijsackers, T. Hammers, (Eds), Kluwer Academic Publishers, 274.

Poenkae A., Salminen E.K. and Ahonen S., 1993, Lead in the ambient air and blood specimen in children in Helsinki, *Sci. Total Environ.,* 138(3): 301-308.

Pope C.A. *et. al.,* 1990 Respiratory Health and PM_{10} pollution: A Daily time series analysis submitted to the American Review of Respiratory Disease.

Pope C.A. III, Schwartz J and Ramson M.R., 1992, Daily mortality and PM_{10} pollution in Utah Valley, *Arch. Environ Hlth.,* 47: 211-217.

Porter J.R. And Lawlor D.W. (Eds), 1991, Plant Growth Interactions With Nutrition And Environment, Society For Experimental Biology, *Seminar Series 43*, Cambridge University Press, New York.

Pouyat R.V., Mcdonel M.J. And Pickett S.T.A., 1995, Soil Characteristics of Oaks Stands Along An Urban-Rural Land Use Gradient, *J. Environ Qual.,* 24: 516-526.

Pratibha, Sharma Madhu, 2000, Changes In Chlorophyll And Free Amino Acids of Gram In Response To Sulphur Di Oxide Exposure Under Field Conditions, *Polln Res.,* 19(1): 95-97.

Przymusinski R., 1980, Cytochemical Localization of Peroxidase Activity In Needles of Scots Pine Resistant And Susceptible To Industrial Pollution In Mejnartowicz L. (Ed), A Genetic Basis For The Resistance of Forest Trees To Anthropopressure With Special Study of The Effect of Some Toxic Gases, Polish Academy of Sciences, Institute of Dendrology Kornik, Poland, 38-49.

Puccinelli Anselmi N., Bragloni M., 1998, Peroxidases: Usable Markers of Air Pollution In Trees From Urban Environments, *Chemosphere* 36 (4-5): 889-894.

Purushothamanan S., Mukundan K., Viswanath S., 1996, *Ind. Journal of Agricultural Economics* 51: 407.

Pye K., 1987, Aeolian Dust And Dust Deposits, Cambridge University Press, Cambridge.

Quality of Urban Air Review Group (QUARG), 1996, Air Borne Particulate Matter In The United Kingdom, *Third Report of The Quality of The Urban Air Review Group, Quarg,* Birmingham.

Querol Xavier, Alastuey Andres, Rodriguez Sergio, Plana Felicia, Ruiz Carmrn R., Cots Nuria, Massague Guillem And Puig Oriol, 2002, Pm10 And Pm2.5 Source Apportionment In The Barcelona Metropolitan Area, Calalonia, Spain, *Atmospheric Environment* 35(36): 6407-6419.

Rabe R. and Kreeb K.H., 1979, Enzyme activities and chlorophyll and protein content in plants as indicators of air pollution, *Environ. Pollut.,* 19: 119-137.

Rabinowitz M.B. *et. al.*, 1988, Effect of food intake and fasting on gastrointestinal lead absorption in humans, *A. J. Clin. Nutr.,* 33:1784-1788.

Rao D. N ., 1989, Department of Botany. Banaras Hindu University, Varanasi – UP Project : Study of Pollution Sink Efficiency, Growth Response Productivity Pattern of Plants With Res To Fly Ash And $SO_{2.}$

Rao D.N., 1971, A Study of the air pollution problem due to coal unloading in Varanasi, India, Proc. Int. Clean Air Congr. 2nd edition (H.M. England, W.T. Breet eds.) New York, Academic Press, 270-273.

Rao D.N., Leblanc F., 1966, Effect of So_2 On Lichen Algae With Special Reference To Chlorophyll, *Bryologist* 69:60-75.

Rao K.S., Gunter R.L., White J.R. And Hosker R.P., 2002, Turbulence And Dispersion Modeling Near Highways, *Atmospheric Environment* 36(27): 4337-4346.

Raskin I. And Ensley B.D., 2000, (Ed.), Phytoremediation of Toxic Metals: Using Plant To Clean Up The Environment, John Wiley And Sons, New York.

Ratan R.K. And Sharma Neelum, 1994, Interaction of Phosphorus With Other Macro And Micro Nutrients, *Proceedings of Group Discussion On Phosphorus Researches In India* Held At IARI, New Delhi, 91-119.

Rawn David J., 1989, Biochemistry, Neil Patterson Publishers, 484-532.

Rathore G.S., Khamparia R.S., Gupta G.P., Dubey S.B., Sharma B.L., And Tomar V.S., 1995, Twenty Five Years of Micronutrient Research In Soils And Crops of Madhya Pradesh, *Research Bulletin*, Dept. of Soil Science And Agricultural Chemistry, J.N. Krishi Vishwavidayala, Jabalpur.

Ray Malabika, 1990, Accumulation of Heavy Metals In Plants Grown In Industrial Areas, *Indian Biologist* XXII(2): 33-38.

Regional Air Quality Concerns, Regional Atmospheric Degradation Emissions Trading To Reduce Sulfur Dioxide Levels Atmospheric Deposition And Impairments of Visibility.

Rennenberg H. And Polle A., 1994, Metabolic Consequences of Atmospheric Sulphur Influx Into Plants In Alscher R. And Welburn A. (Eds.), *Plant Response To Gaseous Environment*, Chapman And Hall, London, 165-180.

Rennerberg H. And Herschblanch C., 1996, Response of Plants To Atmospheric Sulphur In Plant Response To Air Pollution, (Eds.), Yunus. M., Singh.N., Iqbal. M., 285-293.

Reuther W., and Labanauskas C.K., 1966, Copper in Diagnostic criteria for plants and soil, H.D. Chapman (ed.), Abilene, TX, Quality printing, 157-179.

Reynolds A.W. And Broderick B., 2000, Air Pollution And Modeling In Dublin, Air Quality Group, Dept. of Civil, Structural And Environmental Engineering, T.C.D. And Dept. of Mechanical Engineering, N.U.I., Galway.

Rhee D.G., 1991, Air pollution in the megacity of Seoul, Ministry of Environment, Seoul, South Korea

Richardson C.J., Sasek T.W., Di Giulio R.T., 1990, Use of Physiological And Biochemical Markers of Air Pollution Stress In Trees In Wang W., Gorsuch Jw. And Lowe Wr. (Eds.), Plants For Toxicity Assessment, *American Society For Testing And Materials*, Philadelphia, 143-155.

Ristoviski Z.D., Morawska L., Bofinger N.D., Hitchnis J., 1998, Submicrometer and supermicrometer particles from spark ignition vehicles, *Environmental Science and Technology*, 32: 3845-3852.

Roberts. J. Users, Friendly Cities, Test London (1989); Test Quality Streets, Test London (1989); Lutraq The Pedestrian Environment, *Lutraq Report* 4a, 1000 Friends of Oregon Portland (1993).

Rogge W.F., Hildemann L., Mazurek M.A., Cass G.R., Simoneit B.R.T., 1993a, Sources of Fine Organic Aerosol, Non Catalyst And Catalyst Equipped Automobiles And Heavy Duty Diesel Trucks, *Environmental Science And Technology,* 27: 636-651.

Rogge W.F., Hildemann L., Mazurek M.A., Cass G.R., Simoneit B.R.T., 1993b, Sources of Fine Organic Aerosol, Road Dust And Tire Debris And Organometallic Brake Lining Dust: Roads As Sources And Sinks, *Environmental Science And Technology,* 27: 1892-1904.

Rogge W.F., Hildemann L., Mazurek M.A., Cass G.R., Simoneit B.R.T., 1993c, Sources of Fine Organic Aerosol, Natural Gas Home Appliances, *Environmental Science And Technology*, 27: 2700-2711.

Rohn R.D. *et. al.,* 1982, Somatomedin activity before and after chelation therapy in lead intoxicated children, *Arch. Environ. Hlth.,* 37: 369-373.

Romieu I. *et. al.*, 1991, Urban air pollution in Latin America and Caribbean: health perspectives, *J. Air Waste Management Assoc.*

Rovinsky F.V, Egorov V.I, Starodubsky I.A, 1991, Urban Air Quality In Moscow.

Roy Prabal K., And Sukugawa Hiroshi, 2001, Trends of Air Pollution And Its Present Situation In Hiroshima Perfecture, *Water Air And Soil Pollution*, 130: 1805-1810.

Ruusakanen J., Tuch T.H., Ten Brink H., Peters A., Khylstov A., Mrime A., Kos G.P.A., Brunekreef B., Whichmann H.E., Buzorius G., Vallius M., Kreyling W.G., Pekkanen J., 2001, Concentration of Ultra Fine And Pm 2.5 Particles In Three European Cities, *Atmospheric Environment*, 35: 3729-3738.

Saenger P. *et. al.,* 1984, depressed excretion of 6B-hydroxycortisol in lead toxic children, *J. Clin. Endocrin. Metab.,* 58:363-367.

Salmons W., 1993, Non Linear And Delayed Response To Toxic Chemicals In Environment: In Contaminated Soil 93, F. Arendt, G.J. Annokkee, R. Bosman And W.J. Van Den Brink (Eds), Kluwer Academic Publishers, 225.

Salt D.E., Blaylock M., Kumar M., Dushenkev A.N., Ensly V., Chet I Raskini, 1995, Phytoremediation: Novel Strategy For The Removal of Toxic Metals From The Environment Using Plants, *Biol Technol* 13: 468-474.

Samantary S., 2002, Biochemical Responses of Cr-Tolerant And Cr-Sensitive Mung Bean Cultivators Grown On Varying Levels of Chromium, *Chemosphere*, 47(10): 1065-1072.

Samantry S., Rout G.R., Das P., 1999, Studies On The Uptake of Heavy Metals By Various Plant Species On Chromite Mine Spoils In Sub-Tropical Regions of India, *Environmental Monitoring And Assessment*, 55: 389-99.

Samet J.M. and Utell M.J., 1990, The Risk of nitrogen dioxide: What have we learned from epidemiological and clinical studies, *Toxicol. Ind. Hlth.*, 26:247-262.

Sanchez Ccoyllo O.R., Abdrade Fatima De M., 2001, Pollutants Concentration In Sao Paulo Brazil, Departmen of Atmospheric Sciences, Institute of Astronomy And Geophysics University of Sao Paulo, Rua Do Matao 1226, 05508-900.

Santamaria J.M., Martin A., 1998, Influence of Air Pollution On The Nutritional Status of Navarras Forests, Spain, *Chemosphere* 36: 943-948.

Sarkuman V., Mishra A.K., Nayar P.K., 1991, Effect of Compost Lime And Phosphorus On Cd Toxicity In Rice *J. Indian Soc Soil Sci.*, 39(3): 595-597.

Sawadis T., Marnasidis A., Zachariadis G. and Stratits J., 19995, A study of air pollution with heavy metals in Thessaloniki city (Greece) using trees as biological indicators, Arch. Environ. Conntam. *Toxicol.*, 28: 118-124.

SCAG, 1991, Air Quality Management Plan- South Coast Air Basin. South Coast Air Quality Management District, Southern California, Association of Governments Los Angeles.

Schrenk H.H., Heimann, Clayton H.G.D., Gatafer W.M. and Wexler H., 1949, Air Pollution in Donora PA *Public Health Bulletin 306*, US Public Health Service Washington DC.

Schulz Ed., Oren R., Lange O.l., 1989, Processes Leading To Forest Decline: A Synthesis. In Schulze Ed., Oren R., Lange Ol (Eds.) Forest Decline And Air Pollution: A Study of Spruce On Acid Soils. *Ecological Studies* 77: 459-468.

Schulz H., 1986, Festigkeit Und Wassergehalt In Fitchen Kifern Und Buchen Unterschiedlicher Schadstufen, *Holz Als Roh-Und Werkstoff*, 44: 300-301.

Schwar M.J.E, Moorcraft J.S., Laxen D.P.H., Thompson M. And Armorgie C., 1988, Baseline Metal In Dust Concentrations In Greater London, *Science of Total Environment*, 68, 25-43.

Schwartz J and Marcus A., 1990, Mortality and air pollution in London: a time series analysis, *Am. J. Epidemiol.,* 131:185-194.

Schwartz J., 1991, Particulate air pollution and daily mortality, presented at the society for occupational and environmental health. *Health effects of Air Pollution: Impact of clean air legislation*, March 25-27, Crystal city VA, 73.

Schwartz J., Dockery D.W., Neas L.M., 1996, Is Daily mortality associated specifically with fine particles?, *Journal of the Air and Waste Management Associtaion*, 46: 927-939.

Score R.S., 1973, Pollution In The Air, Roultledge And Keeegan Paul, London.

Seaton A., MacNee W., Donaldson K., Godden D., 1995, Particulate air pollution and acute health effects, *The lancet*, 345: 176-178.

Seaward M.R.D. and Richardson D.H.S., 1990, Atmospheric sources of metal pollution and effects on vegetation In Shaw, A.T. (ed.) Heavy Metal tolerance in plants: Evolutionary Aspects, CRC press Inc, Boca Raton, Florida, 75-92.

Sehgal Deepak And Ratan R.K., 1990, Investigation On Cadmium Phyto Toxicity In Peas On Acid And Alkaline Soils, *J. Indian Soc. Soil Sci.* 38(3): 561-564.

SERPLAC, 1989, Regional metropolitan services, Epidemiological study of the effects of atmospheric pollution, Santiago.

Shahare C.B., Varshney C.K., 1994, Impact of Sulphur Di Oxide On Some Trees With Reference To Their Growth, *J. Env. Polln.*, 1(3&4), 149-155.

Sharma G.K., 1977, Cuticular Features As Indicator of Environmental Pollution, *Environ. Pollut.*, 5: 587-593.

Sheehan P.J. *et.al.,* 1991, Assessment of The Human Health Risks Posed By Exposure To Chromium Contaminated Soils, *J, Toxicol. Environ. Health*, 32: 161-201.

Shen S.J., Jaques Peter A., Zhu Yifang, Geller Michael D. And Siotas, 2002, Evaluation of The Smps-Aps System As A Continuous Monitor For Measuring Pm2.5, PM_{10} And Coarse (Pm2.5) Concentrations, *Atmospheric Environment* 36(24): 3939-3950.

Shi J.P., Evans D.E., Khan A.A., Harrison R.M., 2001, Sources and concentration of nonparticles (<10nm diameter) in thr urban atmosphere, *Atmospheric Environment*, 35: 1193-1202.

Sillanpaa M., Jansson H., 1992, Status of Cadmium, Lead, Cobalt And Selenium In Soil And Plants of Thirty Countries *Soil Bulletin 65*, Fao, Rome Italy.

Sindhu R.S., Sharma Reeta, 1999, Bioaccumulation of Lead Cadmium And Mercury By Galium Ticorne, *Pollen Res.*, 18(4): 395-397.

Singh N., Pandey V., Misra J., Yunus M. And Ahmad K.J., 1997, Atmospheric Lead Pollution From Vehicular Emissions-Measurements In Plants, Soil And Milk Samples, *Environmental Monitoring And Assessment* 45: 9-19.

Singh N., Yunus M., Srivastava K., Singh S.N., Pandey V., Misra J., And K.J. Ahmad, 1995, Monitoring of Auto Exhaust Pollution By Road Side Plants, *Environmental Monitoring And Assessment* 34: 13-26.

Singh R.R., Singh Vijay And Shukla A.K., 1991, Yield And Heavy Metal Contain of Berseem As Influenced By Sewage Water And Refinery Effluent, *J. Indian Soc. Soil Sci.* 39(2): 402-404.

Sinha B.P. and Pandey S., 1969, Diesel smoke effecting haemograph, *J. Ind. Med. Ass.,* 55:556-557.

Sinha Suchita, Mukherji S. And Dutta Jayanta, 2002, Effect of Manganese Toxicity On Pigment Content, Hill Activity And Photosynthetic Rate of Vigna Radiata L. *Wilczek Seedlings*, 23(3).

Siversten B., Matale C., And Pereira L.M.R., 1995, Report No. 35. SADC, Environment And Land Management Sector Coordination Unit, Pretoria, South Africa.

Smith W.H, 1981b, Air Pollution And Forests, Springer Verlag, New York.

Smith W.H., 1981a, Air Pollution And Forest Interaction Between Air Contaminants And Forest Ecosystem Springer Verlag, New York.

Solane C.S. and Tesche T.W., 1991, Atmospheric chemistry: models and prediction for climate and air quality, Lewis Publisher, Chelsea, MI.

Solberg R.A. And Adams D.F., 1950, Histological Responses of Some Plant Species To Hf And So_2, *Amrican J. Bot.*, 43: 755-766.

Southern Oxidant Study Report, 1990, University Corporation for Atmospheric Research, August.

Sriramachari S. and Harish Chandra, 1997, The lessions of Bhopal (toxic) MIC gas disaster scope for expanding global biomonitoring, *Chemosphere*, 34: 2237-2250.

Srivastava A.K., Azhar S. And Krishnamurthy C.R., 1972, Inhibition of Germination In Cicer Arietinum, *Phyotochemistry*, 11: 3181-3185.

Stedman John R., 2002, The Use of Receptor Modeling And Emission Inventory Data To Explain The Downward Trend In U.K. PM_{10} Concentrations, *Atmospheric Environment* 36(25): 4089-4101.

Steenland K, 1986, Lung Cancer and Diesel Exhaust: A Review, *Am. J. Ind. Med.,* 10:177-189.

Stefanov K. *et. al.*, 1992, Lipid And Sterol Changes In Phaseolous Vulgaris Caused By Lead Ions, *Phytochemistry*, 31: 3745-3748.

Stefanov K. *et. al.*, 1993, Lipid And Sterol Changes In Zea Mays Caused By Lead Ions, *Phytochemistry*, 33: 47-51.

Stefanov K. *et. al.*, 1995, Effect of Lead Ions On Lipids And Antioxidant Complex Activity of Capsicum Annum L. Leaves, Pericarp And Seeds, *J. Sci. Food. Agric.,* 67: 259-266.

Stern F. *et. al.*, 1988, Heart disease mortality among bridge and tunnel officers exposed to CO, *Am. J. Epidemiol.,* 128:1276-1288.

Stocker G. And Gluch W., 1990, Bioindication of Acid Deposition On Forest Ecosystem- Recognition of Local And Regional Pattern, Arch. Nat. Schultz. Landsch. *Forsch.*, 30: 3-12.

Strain H.H., Benganvin T.C. and Walter A.S., 1971, Analytical procedure for isolation, identification, estimation, investigation of chlorophyll In *Methods in Enzymology* (A.S. Pietro ed.) New York, 23: 452-476.

Street R.A., Duckham S.C., Hewitt C.N., 1996, Laboratory and field studies of volatile organic compound emissions from Sitka Spruce (Picea sitchensis Bong.) in the UK., *J. Geophysical Research Atmospheres*, 101: 22799-22806.

Streit B, And Stumm W, 1993, Chemical Properties of Metals And The Process of Bioaccumulation In Terrestrial Plants In Market (Ed) Plants As Biomonitors, Indicators of Heavy Metals In Terrestrial Environment (Chapter 2) Weinheim: Vhc 31-62.

Stroup N. *et. al.*, 1984, Brain Cancer and other causes of death in anatomists, *Am. J. Epidemiol,* 120:500.

Sturaro A., Parvoli G., Doretti L., 1993, Plane Tree Bark As A Passive Sampler of Polycyclic Aromatic Hydrocarbons In Am Urban Environment, *Journal of Chromatography* 643: 435-438.

Takada H., Onda T. and Ogura N., 1990, Determination of PAHs in urban street dusts and their source materials by capillary gas chromatography, *Environ. Sci. Technol.,* 24:1179-1186.

Takeuchi Y., Mabuchi C. and Takagi S., 1975, Polyneuropathy caused by petroleum benzene, *Int. Arch. Arbeitsmed,* 34:185-190.

Tang X., Li. I., Chen D., Bai Y., Li., X., Wu X., and Chen J., 1988, the study of photochemical smog pollution in China, In *Proceedings third joint conference of air pollution studies in Asia*, 238-251, Japan society of air pollution Tokyo, Japan.

Thomson J.R., Mueller P.W., Fluckiger W., Rutter A.J., 1984, The Effect of Dust On Photosynthesis And Its Significance For Roadside Plants, *Environ. Pollut.,* Series A, 34, 171-190.

Times of India, January, 12, 2001, www.Timesofindia.Com/120101/12indi28.html.

Titta Petri, Raunemaa Taisto, Tissari Jarkko, Yli-Tuomi Tarja, Leskinen Ari, Kukkonen Jaakko, Harkonen Jari And Karppinen Ari, 2002, Measurements Anf Modeling of Pm2.5 Concentrations Near A Major Road In Kuopio, Finland, *Atmospheric Environment* 36(24): 3973-3988.

Tiwari S., Bansal S., 1993a, Effect of NO_2 Pollution On Mimusops Ellengi Linn., *Asian J Plant Sci.,* 5(1), 83-87.

Tiwari S., Bansal S.,1993b, Assessment of Air Pollution Tolerance of Two Common Tree Species Against SO_2, *Biome*, 6(2), 78-82.

TMG, 1989a, Automobile pollution control plan: Towards to a better living program in Tokyo, Tokyo metropolitan Government, Tokyo.

TMG, 1989b, Summery of results of monitoring of air pollution in the FY 1989, Air monitoring division Tokyo.

Tringey D.T. And Reinert R.A, 1975, The Effect of O_3 And SO_2 Singly And In Combination On Plant Growth, *Environ. Pollut.*, 9: 117-125.

Tripathi Ashutosh Kumar, Tripathi Sadhna, 1999, Changes In Some Physiological And Biochemical Characters In Albizia Lrbbek As Bioindicators of Heavy Metal Toxicity, *J. Environ. Bio.,* 20(2): 93-98.

Tschalinski T.J. And Norby R.J., 1991, Physiological Indicators of Nitrogen Response In Short Rotation Sycamore Plantation, I. CO_2 Assimilation Photosynthetic Pigments And Soluble Carbohydrates, *Physiol. Plant.*, 82: 117-126.

Tsukahara H., Kozlowski T.T., Shanklin J., 1985, Tolerance of Pinus Densiflora, Pinus Thunbergii And Larix Leptolepis Seedlings To SO_2, *Plant And Soil*, 88: 385-398.

Tuppurainen M. *et. al.*, 1988, Thyroid function as assessed by routine laboratory tests of workers with long term exposure, *Scand. J. Work Environ. Hlth.,* 14:175-180.

UK-TERG, 1988, Interactions Between Air Pollutants And Other Factors, In Effects of Acid Deposition On The Terrestrial Environment In The Uk, HMSO, London: 73-83.

UNEP/GEMS, 1991, Urban Air Pollution, United Nations Environment Program, Oxford.

United States Environment Protection Agency (USEPA), March 2000, National Air Pollutants Emission Trends, 1900-1998, Epa-454/R-00-002, Office of Air Quality Planning And Standards, Research Triangle Park, N.W.

USEPA, 1982,Air quality criteria for particulate matter and sulphur oxides, Office of research and development, Environment criteria and assessment office, Research Triangle Part, NC.

USEPA, 1991, National Air Quality And Emissions Trend Report 1989 USEPA, Research Triangle Park, N.C.

Vahter M. And Sloarch S., 1990, Exposure Monitoring of Lead And Cadmium: An International Pilot Study With In The WHO/UNEP, Human Exposure Assessment Location (Heal) Programme, UNEP Nairobi.

Vaishampayan. J.V., 1996, Sustainable Development And Quality of Life, Center For Adult, Continuing Education, Lucknow University.

Vardarjan S., Doraiswamy L.K., Ayyangar N.R., Iyer C.S.P., Khan A.A., Lahiri A.K., Mazumdar K.V., Mashelkar R.A., Mitra R.B., Nambear O.G.G., Ramchandaran V., Shastrabudhe V.D., Sivram S., Sraram S., Thycegrajan G. and Venketraman R.S., 1995, Report on scientific studies on factor related to Bhopal toxic gas leakage, CSIR, New Delhi.

Varshney C.K., Agarwal M., 1997, *Atmos. Environ*. 26b: 291

Vaughan T.L. *et. al.*, 1986, Formaldehyde and cancers of the *pharynx, sinus*, and nasal cavity II. Residential exposures, *Int.J. Can.,* 38:685-688.

Vernon L.P., 1960, Spectrophotometric determination of Chlorophylls and Phaeophytin in plant extract, *Anal. Chem*., 32: 1144-1150.

Voet Donald and Voet Judith G., 1990, Biochemistry, John Wiley & Sons, 586-617.

Wahid A., Milne E., Marshall F.M., Shamsi S.R.A., Ashmore M.R. And Bell J.N.B., 1993, Final Report, U.K. Deptt. For International Development Research Project No. 206-N Punjab University Lahore And Imperial College, London

Waleed H.A., Motaberuddin A., Khalis M.M., 1990, Measurement of ambient air lead concentration in the city of Jeddah, Saudi Arabia, *Environ. Int.,* 16: 85-88.

Walrath J. and Fraumeni J.F., 1983, Cancer and other causes of death among embalmers, *Can. Res.,* 44:4638-4641.

Wards N., Brooks R.R. and Roberts E., 1977, Heavy metal pollution from automotive emissions and its effect on roadside soil and pastuer species in New Zealand, *Environ. Sci. Technol..,* 11: 917-921.

Ware J.H. et. al., 1986, Effects of ambient sulphur oxides and suspended particles on respiratory health of preadolescent children, *Am. Rev. Resp. Dis.,* 133:834-842.

Warneck P., 1988, Chemistry of Natural Atmosphere, Academic Press, San Diego.

Weber R., And Hrynczuk B., 2000, Effect of Leaf And Soil Contamination On Heavy Metal Content In Spring Wheat Crops, Institute of Soil Science And Plant Cultivation, Lakowa 2 Str., 55-230 Jelcz-Laskowice, Poland.

Welburn A.R., 1990, Air Pollution And Acid Rain: The Biological Impact. Longman Scientific And Technical, London.

Wenzel W.W., and Jockwer F., 1999, Accumulation of heavy metals in plants grown on mineralized soil of the Australian Alps, *Environ. Pollut.,* 104: 145-155.

Whelphale D.M., 1992, An Overview of The Atmospheric Sulphur Cycle In Haworth Rw., Stewart Jwb. And Ivanov Mv., (Ed.) Sulphur Cycling On The Continents, Scientific Committee On Problems of Environment 48, Chicester, Uk, 5-26.

Whitmore M.E. And Mannsfield T.A., 1983, Effect of Long Term Exposure To So_2 And No_x On Poa Pratensis And Other Grasses, *Environ. Pollut.,* 31: 317-235.

WHO/UNEP, 1992, Urban Air Pollution In Mega Cities of World. Cambridge, MA. Blackswell Publishers, Oxford.

WHO/UNEP/GEMS, 1998 Assessment of Urban Air Quality of Geneva.

Wichmann H.E., Muller W., Allhoj P. *et. al.*, 1989, Health effects during a smog episode in East Germany in 1985, *Environ. Hlth. Perspect.,* 79:89-99.

Wilfried H.O. Ernst, 1999, The Role of Sulphur In Responses of Plants To A Surplus of Heavy Metals, Deptt. of Ecology And Ecotoxicology of Plants, Faculty of Biology, Vrije University, De Boelelaan 1087, 1081 HV Amsterdam, Netherlands, www.meurei.hpg.ig.com.

Williams E., 1995, Inter-comparison study: southern oxidants study 1995, data analysis workshop and annual meeting Raleigh, NC.

Wilkinis E.T., 1954, Q.Jr., *Meteorol. Soc.,* 80:267-271.

Wilkinis E.T., 1994, Jr., *Sanit Inst.,* 74:1-15.

Wilson G.B. And Bell J.N.B., 1985, Studies On The Tolerance To SO_2 of Grass Populations In Polluted Areas L[11], Investigations On The Rate of Development of Tolerance, *New Phytol.*, 100: 63-77.

Witting R., 1993, General aspects of biomonitoring heavy metals by plants In B. Markert (ed.), Plants as Biomonitors: Indicators for heavy metals in the terrestrial environments, VCH, Weineim, 3-28.

Wolfenden J., Pearson M. And Francis J., 1991, Effects of Over Winter Fumigation With Sulphur And Nitrogen Oxides On Biochemical Parameters And Spring Growth In Red Spruce (Picea Rubens), *Plant Cell Environ.,* 14: 35-45.

Wolman. A., The Metabolism of The City (1965), 1992, Scientific American 213, 179, Boyden. S., Millar. S., Newcombe. K., And O' Neil L. B., The Ecology of A City And Its People, Anu Press, Canberra (1981), Girardet. H., The Gaia Atlas of Cities, London.

Wolterbeek Bert, 2002, Biomonitoring of Trace Element Air Pollution: Principles, Possibilities And Perspectives, *Environmental Pollution* 120(1): 11-21.

Wong J.W.C., Lai K.M., Su D.S., Fang M., 2001, Availability of Heavy Metals For Brassica Chinensis Grown In An Acidic Loamy Soil Amended With A Domestic And An Industrial Sewage Sludge, *Water Air And Soil Pollution,* 128: 339-353.

Wong S.C., Lin X.D., Zhang G., Min Y.S., 2002, Heavy Metals In Agricultural Soils of The Pearl River Delta, South China, *Environmental Pollution* 119(1): 33-44.

Woolhouse H.,1986, Plant Senescence In: Process And Control of Plant Senescence (Eds), Leshem Yy, Halvey A.W., Frankel C., Elsevier Science Publishers, Nethenlands.

World Health Organization, 1977, NO_X, Geneva, *Environmental Health Criteria* No. 4.

World Health Organization, Working Group, 1977, Lead, Geneva, *Environmental Heath Criteria* No. 3:1-60.

World Health Organization, 1978, Photochemical Oxidants, Geneva, *Environ. Health Criteria* No. 7.

World Health Organization, 1979, CO, Geneva, *Environmental Health Criteria* No. 13.

World Health Organization, 1987a, Nitrogen di oxide In Air quality guidelines for Europe, Copenhagen, WHO regional office for Europe (*WHO Regional Publications*), European Series No. 23.

World Health Organization, 1987b, Ozone and other photochemical oxidants In Air quality guidelines for Europe, Copenhagen, WHO regional office for Europe (*WHO Regional Publications*), European Series No. 23.

World Health Organization, 1987c, Sulphur dioxide and particulate matter In Air quality guidelines for Europe, Copenhagen, WHO regional office for Europe, *(WHO Regional Publications), European Series* No. 23.

World Health Organization, 1987d, Carbon monoxide In Air quality guidelines for Europe, Copenhagen, WHO Regional Office for Europe *(WHO Regional Publications*), European Series No. 23.

World Health Organization, 1987e, Lead In Air quality guidelines for Europe, Copenhagen, WHO regional office for Europe (*WHO Regional Publications*), European Series No. 23.

World Health Organization, 1987f, Benzene In Air quality guidelines for Europe, Copenhagen, WHO Regional Office for Europe, *WHO Regional Publications*, European Series No. 23.

World Health Organization, 1987g, Poly aromatic Hydrocarbons (PAHs) In Air quality guidelines for Europe, Copenhagen, WHO regional office for Europe (*WHO Regional Publications*), European Series No. 23.

World Health Organization, 1989, Formaldehyde, Geneva, *Environment Health Criteria No.* 89.

WSL, 1992, Initial Analysis of NO_2 Pollution Episode, December 1991, Warren Spring Laboratory, Stevenage, UK.

Xian X, 1990, Relationship Between Distribution of Cadmium In Root Tissue And Cd- Tolerance of Athyrium Yokoscense, *J. Environ. Sci. Health,* A25(6): 621-628.

Xiong Z. 1998, Heavy metal contamination of urban soils and plants in relation to traffic in Wuhan city, China, Toxicological and Environmental Chemistry, 65: 31-39.

Xizn X., 1990, Relationship between distribution of Cadmium in Root Tissue and Cd-tolerance of Athyrium Yokoscense, *J. Environ. Sci. Hlth.,* 25(6A): 621.628.

Yassoglou N., Kosmas C., Asimakopoulos J. and Kallianou C., 1987, Heavy Metal contamination of roadside soil in the grater Athens area, *Environ. Pollut.,* 47: 293-304.

Yin S.N. *et. al.,* 1989, A Retrospective study of leukemia and other Cancers in benzene workers, *Environ. Hlth. Perspect.*, 82:207-213.

Yunus Modd. And Ahmad K.J. And Gale R., 1979, Air Pollutants and Epidermal Traits In *Ricinus Communis L., Environ Pollut.* 20: 189-198.

Yunus. M., Singh.N., Iqbal. M., 1996, Global Status of Air Pollution: An Overview; Plant Response To Air Pollution, (Eds).

Zhang Xiao Y., Caoj.J., Li L.M., Arimoto R., Cheng Y., Huebert B., And Wang D., 2002 Characterization of Atmospheric Aerosol Over Xian In The South Margin of The Loess Plateau, China, *Atmospheric Environment*, 36(26): 4189-4199.

Zheng Y. And Chen S., 1991, Atmospheric Environment (China) 6: 45 (In Chinese)

Zhu Yifang, Hinds William C.,Kim Seongheon, Shen Si And Sioutas Constantinos, 2002, Study of Ultrafine Particles Near A,Ajor Highway With Heavy Duty Diesel Traffic, *Atmospheric Environment* 36(27): 4323-4335.

Zhu Yifang, William C. Hinds, Seongheon Kimn And Constantinos Sioutas, 2002, Concentration And Size Distribution of Ultrafine Particles Near A Major High Way, Technical Paper, *Journal of The Air And Waste Management Association*, 52: 1032-1042.

Ziegler I., 1975, *Res. Rev.,* 56: 79.

Ziegler E.E. *et. al.,* 1978, Absorption and retention of lead by infants, *Pediatric Res.,* 12:29-34.

Zimmermann R., Oren R., Schulze Ed., Werk Ks., 1988, Performance of Two Abies Stand (L) Karst At Different Stages of Decline. in. Photosynthesis And Leaf Conductance, *Oecologia* 76: 513-518.

Zummo S.M. And Karol M.H., 1996, Indoor Air Pollution- Acute Adverse Health Effects And Host Susceptibility, *J. Environ. Hlth.,* 58: 25-29.

Index

Zeitfracht Medien GmbH
Ferdinand-Jühlke-Straße 7
99095 Erfurt, Deutschland
produktsicherheit@kolibri360.de